W9-DAY-360

EMBODIED GEOGRAPHIES

The life-course can be seen as a journey, and this book charts our progress, revealing how we cope with the rough passages: the crisis points that can be termed 'rites of passage'. Life crises involve learning about our bodies, about the places we live in, and about our identity. They challenge us to draw on our reserves of adaptability and can transform how we live our lives and how we see ourselves.

Embodied Geographies provides an account of different types of life moments and stages which can contribute to forging our identities. Chapters focus on pregnancy, childbirth, parenthood, threat and reality of violence, illness, disability, migration, bereavement and the ensuing changes to family relationship and responsibilities, and death itself. Within these accounts, we are made aware of how, over the life-course, our horizons widen during childhood but close in towards the end of our lives; how we learn to stake claims to terrain during the rites of the school day or the celebrations of school-leaving; how new parents struggle with deeply embedded patriarchal norms to negotiate their everyday time–space routines; what 'ableist space' means to the physically impaired, the ill or victims of violence; how migrants weld new and former homes into their sense of identity. Each chapter examines aspects of the close relationship between body, place and identity. The geographical spread of the case studies enables comparisons to be made between how people cope with crucial life transitions in six of the world's wealthiest nations: Australia, New Zealand, Great Britain, Singapore, the USA and Canada, and also in Hong Kong.

Elizabeth Kenworthy Teather is a Senior Lecturer in Geography and Planning at the University of New England, Australia.

CRITICAL GEOGRAPHIES

Edited by Tracey Skelton

Lecturer in International Studies, Nottingham Trent University

and

Gill Valentine

Senior Lecturer in Geography, The University of Sheffield.

This series offers cutting-edge research organised into three themes: concepts, scale and transformation. It is aimed at upper-level undergraduates and research students, and will facilitate inter-disciplinary engagement between geography and other social sciences. It provides a forum for the innovative and vibrant debates which span the broad spectrum of this discipline.

1. MIND AND BODY SPACES
Geographies of illness, impairment and disability
Edited by Ruth Butler and Hester Parr

2. EMBODIED GEOGRAPHIES
Spaces, bodies and rites of passage
Edited by Elizabeth Kenworthy Teather

3. LEISURE/TOURISM GEOGRAPHIES
Practices and geographical knowledge
Edited by David Crouch

4. CLUBBING
Dancing, ecstasy, vitality
Ben Malbon

5. ENTANGLEMENTS OF POWER
Geographies of domination/resistance
Edited by Joanne Sharp, Paul Routledge, Chris Philo and Ronan Paddison

6. DE-CENTRING SEXUALITIES
Politics and representations beyond the metropolis
Edited by Richard Phillips, Diane Watt and David Shuttleton

EMBODIED GEOGRAPHIES

Spaces, bodies and rites of passage

Edited by
Elizabeth Kenworthy Teather

London and New York

First published 1999
by Routledge
11 New Fetter Lane, London EC4P 4EE

Simultaneously published in the USA and Canada
by Routledge
29 West 35th Street, New York, NY 10001

Routledge is an imprint of the Taylor & Francis Group

Typeset in Perpetua by The Florence Group, Stoodleigh, Devon
Printed and bound in Great Britain by Biddles Ltd, Guildford and King's Lynn

British Library Cataloguing in Publication Data
A catalogue record for this book is available from the British Library.

Library of Congress Cataloging in Publication Data
Embodied geographies: spaces, bodies and rites of passage/edited by Elizabeth Kenworthy Teather.
p cm. – (Critical geographies)
1. Identity (Psychology) 2. Human geography. 3. Rites and ceremonies. 4. Personal space.
I. Teather, Elizabeth K. (Elizabeth Kenworthy). II. Series.
BF697.E55 1999
155.9–dc21

ISBN 0–415–18439–8 (hbk)
ISBN 0–415–18440–1 (pbk)

CONTENTS

List of tables vii
List of figures viii
Preface ix
Acknowledgements x
Notes on contributors xi

1 **Introduction: geographies of personal discovery** 1
ELIZABETH KENWORTHY TEATHER (AUSTRALIA)

2 **The expanding worlds of middle childhood** 27
MARGARET JONES AND CHRIS CUNNINGHAM (AUSTRALIA)

3 **Messages about adolescent identity: coded and contested spaces in a New York City high school** 43
KIRA KRENICHYN (USA)

4 **Schoolies Week as a rite of passage: a study of celebration and control** 59
HILARY P.M. WINCHESTER, PAULINE M. McGUIRK AND KATHRYN EVERETT (AUSTRALIA)

5 **Pregnant bodies, public scrutiny: 'giving' advice to pregnant women** 78
ROBYN LONGHURST (NEW ZEALAND)

6 **Bodily speaking: spaces and experiences of childbirth** 91
SCOTT SHARPE (AUSTRALIA)

7 **Putting parents in their place: child-rearing rites and gender politics** 104
STUART C. AITKEN (USA)

8 **Women's experiences of violence over the life-course** 126
RACHEL PAIN (UK)

9 **Life at the margins: disabled women's explorations of ableist spaces** 142
VERA CHOUINARD (CANADA)

10 **Journeying through M.E.: identity, the body and women with chronic illness** 157
PAMELA MOSS AND ISABEL DYCK (CANADA)

11 **Identity and home in the migratory experience of recent Hong Kong Chinese-Canadian migrants** 175
WENDY W.Y. CHAN (HONG KONG)

12 **Embodying old age** 193
RICHARD HUGMAN (AUSTRALIA)

13 **The transition into eldercare: an uncelebrated passage** 208
BONNIE C. HALLMAN (CANADA)

14 **Singapore's widows and widowers: back to the heart of the family** 224
PEGGY TEO (SINGAPORE)

15 **The body after death: place, tradition and the nation-state in Singapore** 240
BRENDA S.A. YEOH (SINGAPORE)

Name index 257
Subject index 263

TABLES

2.1 Play ranges of Australian children 34
2.2 How Lismore children spent their leisure time on the day prior to survey 36
2.3 Location of images of play places 36
2.4 Content of children's photographs 37
2.5 People in photographs 37
6.1 Dualisms that characterise the politics of childbirth 95
7.1 Time budget representation of John and Lisa's day before interview 119
10.1 Socio-demographic profiles of women diagnosed with M.E. with year of diagnosis, Victoria region, British Columbia, Canada 165
10.2 Descriptions of employment and income 167
13.1 Average hours per week of eldercare, by time-distance, care-giver gender and responsibility for children 214
13.2 Average time-distances travelled to provide care in four tasks, by assistance frequency 216
13.3 Residential relocation of the elderly relative, by care-giver gender 217
13.4 Residential relocation of the elderly relative, by time-distance and care-giver gender 218
14.1 Type of dwelling of the elderly and of all Singaporeans 229
14.2 Educational level of the widowed 229
14.3 Head of household in the residence and tenancy status of the residence 230
14.4 Household membership in the residence 231

FIGURES

4.1	Location of the Gold Coast and Surfers Paradise, Queensland, Australia	62
7.1	The scale of parenting	107
7.2	Sources of guidance for first-time parents (a) perceptions prior to birth; (b) perceptions one year after birth	110
7.3	Importance of elder parents as example (a) perceptions prior to birth; (b) perceptions one year after birth	111
7.4	Perceptions of disruptions to free time (a) prior to birth; (b) perceptions one year after birth	116
7.5	Perceptions of disruptions to employment: (a) prior to birth; (b) perceptions one year after birth	117
15.1	Chinese cemetery at the Choa Chu Kang complex	245
15.2	Columbarium block at the Mount Vernon complex	246

PREFACE

It was during Scott Sharpe's conference presentation at the 28th International Geographical Congress in The Hague, The Netherlands, in August 1996, that I had the idea of putting a book together on the theme of the geographies of rites of passage. Scott's paper was on issues to do with birth and Birth Centres. At the time, I was about to start my second year as Scholar in Residence in the David C. Lam Institute of East-West Studies, at Hong Kong Baptist University. Somewhat to my surprise, I had found myself embarked upon a research project that concerned the worlds of custom and belief behind the powerful landscapes of Chinese cemeteries and graves in Hong Kong. When the conference session finished I chatted with Scott about the idea of a book on the geographies of rites of passage.

A little later in the Congress, I arranged with Brenda Yeoh to spend a short period at the National University of Singapore so that we could get together about our mutual research interests into Chinese landscapes of death. While there, I met Peggy Teo, who had research interests in widowhood. In the back of a car on a hot, wet afternoon, we discussed who might be appropriate contributors.

Later that year, when the idea was still at the back of my mind, Jo Campling came to Hong Kong Baptist University to run a week's workshops and consultations on book publishing. Following her guidelines, I prepared a submission and gradually collected together a team of people who wanted to contribute. On my return from Leave of Absence to a full complement of teaching commitments at the University of New England, I was greeted with an acceptance from Routledge. However, the ground had been laid, and it was from then on a question of putting the plans into effect. It has been a delight to work with the contributors on completing the project.

Elizabeth Kenworthy Teather
Armidale, August 1998

ACKNOWLEDGEMENTS

My close colleague during the preparation of this book has been Jan Hayden. Her expertise and meticulous approach to the editing process, combined with her unfailing enthusiasm and endless patience, have been a constant 'calm eye in the storm'. Contributors have been good-naturedly efficient in responding to many e-mail messages. My colleagues in Geography and Planning at the University of New England have tolerated my usurpation of our limited secretarial assistance with forbearance, for which I am grateful. As ever, David Teather has been a source of inspiration and encouragement. I should also like to thank former colleagues from the Department of Education, Hong Kong Baptist University, particularly Atara Sivan, Rebecca Lam and Vicky Tam, and also Brenda Yeoh and Peggy Teo from the Department of Geography, National University of Singapore, all of whom, in its early stages, were supportive of the idea of this book.

The poem *Seashells and Sandalwood* is reproduced from *A Counterfeit Silence* (1969) by Randolph Stow, with the permission of the publishers, Angus and Robertson. Rhoda Howard has given permission for us to publish her two poems, *My Reality* and *Transformation*. We acknowledge this with thanks. Emily Brontë's poem *And Like Myself Lone, Wholly Lone*, is reproduced by permission of Oxford University Press.

NOTES ON CONTRIBUTORS

Stuart C. Aitken is Professor of Geography at San Diego State University, California, USA. He is co-editor of *The Professional Geographer*. His research interests include children, families and film.

Wendy W.Y. Chan is Assistant Director of the David C. Lam Institute for East-West Studies, Hong Kong Baptist University. She is a graduate of the University of Toronto and of the Hong Kong University of Science and Technology. Her academic interests lie in international migration and in higher education policy and management, including the internationalisation of higher education.

Vera Chouinard is Professor of Geography at McMaster University, Hamilton, Ontario, Canada. She specialises in political, economic and feminist approaches to the state, law and urban development. She is also involved, together with those affected, in research into the struggle for co-operative housing, community-based legal services, and disabled persons' access to and experiences of paid and voluntary work

Chris Cunningham is Associate Professor in Geography and Planning at the University of New England, Armidale, New South Wales, Australia. He is an architect and planner. His research interests include children's play, bushfires and community response to them, national myths and legends.

Isabel Dyck is Associate Professor in the School of Rehabilitation Sciences, University of British Columbia, Canada. Her research interests are in the area of women's health, women and suburbs, and feminist methodology with a particular focus on domestic and paid labour issues.

Kathryn Everett is a recent Honours graduate from, and now research assistant in, the Department of Geography and Environmental Science, University of Newcastle, New South Wales, Australia.

Bonnie C. Hallman is Assistant Professor in Geography and Planning at California State University, Chico. She completed her doctorate in 1997. In it she examined the relationship between space/place context and the balancing of work and eldercare commitments by employed Canadian men and women. She is developing new research areas into rural ageing, health and social services provision, comparing Ontario and Northern California.

Richard Hugman is Professor and Head of the School of Social Work at Curtin University of Technology, Western Australia. He has practised in the field of care for older people, as well as researching a range of topics concerning later life.

Margaret Jones is a graduate of the University of New England with particular interest in children. She has worked for some years with Chris Cunningham researching children's play and the associated environmental opportunities.

Kira Krenichyn is a graduate student at the City University of New York Graduate School and University Centre. She has researched the lives of children and adolescents in New York City's public housing and schools.

Robyn Longhurst is Senior Lecturer in the Department of Geography at the University of Waikato, Aotearoa/New Zealand. She teaches on gendered spaces, the body, critical social theory and qualitative methods.

Pauline M. McGuirk is Lecturer in the Department of Geography and Environmental Science, University of Newcastle, New South Wales, Australia. Her main research interests are in urban development, entrepreneurial urban planning and local governance.

Pamela Moss is Associate Professor in the Faculty of Human and Social Development at the University of Victoria, British Columbia, Canada. She teaches as a feminist in research methods and geographies of women. In addition to constructing a radical body politics through drawing on women's experience of chronic illness, her current research involves intersections among autobiographies and concrete spatialities.

Rachel Pain is Senior Lecturer in Human Geography in the Division of Geography and Environmental Management, University of Northumbria at Newcastle, UK. Her main research interests are in urban social geography, particularly geographies of crime, fear of crime, gender and ageing.

Scott Sharpe is a postgraduate student in Human Geography at Macquarie University, Sydney, Australia. His doctoral thesis is concerned with the interrelations between space, subjectivity and academic knowledge production.

Elizabeth Kenworthy Teather is Senior Lecturer in Geography and Planning at the University of New England, Armidale, New South Wales, Australia. Her research areas include rurality, rural social sustainability, rural women's organisations, and the concept of collective memory. She is currently researching contemporary Chinese beliefs relating to death, and the associated landscapes, in Hong Kong.

Peggy Teo is Senior Lecturer in the Department of Geography at the National University of Singapore. Her research interests include ageing policies in the Asia-Pacific region; the socio-spatial environments encountered by older people; and the gendered dimensions of ageing. She is also involved in tourism research.

Hilary P.M. Winchester is Professor and President of Academic Senate, University of Newcastle, New South Wales, Australia. She is a joint editor of *Australian Geographical Studies*, and a member of the Social Sciences Panel of the Australian Research Council. Her main research interests are in urban social geography, gender, and marginal groups.

Brenda S.A. Yeoh is Associate Professor in the Department of Geography at the National University of Singapore, where she is also Director of the Centre for Advanced Studies. Her research and teaching interests include historical and social geography, and the geography of gender. She is editor of the *Singapore Journal of Tropical Geography*.

1

INTRODUCTION: GEOGRAPHIES OF PERSONAL DISCOVERY

Elizabeth Kenworthy Teather (Australia)

> The life of an individual in any society is a series of passages from one age to another and from one occupation to another.
>
> (Arnold van Gennep 1960: 2–3)

The discipline of geography as we know it in the academy today began as a mapping project – to get to know and record the features of the earth's surface. Explorers and navigators brought back new maps and amazing stories from their journeys into unknown lands – journeys often undertaken with an imperialist, mercantile or scientific project in mind. In this book, we examine another sort of journey, equally rich in intriguing experiences and encounters, and undertaken by each one of us. It is a journey through time and space, from birth to death: a journey of personal discovery, during which periods of calm weather are interrupted by more tumultuous passages. It is the geographies of such passages, or *life crises* (Kimball 1960: vii), that are the focus of this book.

In using the term *rites of passage* to refer to these life crises, we concentrate on the term *passage*. We interpret *passage* in the sense of *transition*.[1] Our passages, intensely personal, thread their way through, impact upon and are influenced by the institutional fabric of social life: home, work, school, family, religion, nation, for example. They make huge demands on our personal skills of navigation, as our chapters will show.

By focusing on those periods in an individual's life that are characterised by brief or prolonged periods of significant personal change, we converge, as academics, with creative artists, who find the personal turmoil involved at such turning points to be the driving impetus for creative expression. Some authors in this volume are either writing from an autobiographical point of view or, in basing their chapters on interviews and attending closely to the nature of their subjects' experiences, are '(co)creators of storied knowledge(s)' (Kearns 1997: 269). Thus, there is somewhat less of the 'distance from the topic' that has been traditional in geographical academic writing. The geographies that we present are 'radical and emancipatory' in that our approach has been one of 'getting close to other people, listening to them, making way for them' (Sibley 1995: 184).

In the rest of this introduction, I want first to set the chapters that follow in the context of the academic debate within the discipline of geography about space, place and the body. Second, I will discuss the idea of *rites of passage*. Finally, after introducing each chapter, I will draw out some emerging themes.

The debate about space and place

> Place has to be one of the most multi-layered and multipurpose keywords in our language.
>
> (Harvey 1996: 208)

Concepts of space and place are highly complex, fascinating and much debated. In this section I will begin with the more straightforward conceptual approaches and then move on to some of the less obvious ways in which we encounter, and conceptualise, space and place.

First, as we move from one location to another in the course of the day, we experience places as material, bounded localities. This is the *space of place* (Walmsley 1988). Such specific places acquire meaning for us, in many ways which phenomenologists have tried to tease out (Buttimer 1980) and which are described in countless autobiographical and fictional accounts. The term *sense of place* refers to this link between place and meaning – an existential quality, difficult to define, sometimes shared by many, sometimes different for each individual. *Sense of place* sums up the unique character of some of the places that are part of our lives. We grow attached to such places – whether they are in the city or the country – are defensive about them when they are threatened, and feel bereft when they are destroyed (Tuan 1974; Relph 1976; Porteous 1989; Teather 1990, 1998; Read 1996). In a sense, such places become part of us – of our identity. We regard the most special of such places as *home* – an ideal melding of place, culture and beloved people.

Undoubtedly, increased personal mobility has reduced the rich, varied, small-scale texture of place that we can expect to encounter on a daily basis. Our enhanced mobility is reflected in the concept of *activity space*. Many city dwellers are no longer reliant on the proximity of friends in the next street, but live in 'communities without propinquity' (Webber 1963) i.e. networks of companions or contacts who are physically far-flung. Activity spaces comprise our many *communities of interest*, such as political lobby groups or recreational societies. Activity space is primarily a conceptual, but partly a material, space, as our interactions often literally 'take place'. However, with the development of information technology, networks like those mentioned are stretching our personal experience of time and space across the globe. Cyberspace is an increasingly important 'virtual space' where people 'meet'.

Castells (1989) calls the activity spaces created and maintained by informational technology the *space of flows* (this term is also used by Lefebvre 1991: 354). He argues that the emergence of the space of flows is a 'subtle . . . form of social disintegration and reintegration'. However, the resultant 'meaninglessness of places, the powerlessness of political institutions are resented and resisted, individually and collectively by a variety of social actors'. Such resistance to the insidious power exercised through the space of flows, he argues, needs to be based in identifiable places at the sub-national level: cities and regions (Castells 1989: 348–353). His is a vision of people consciously co-operating in bounded places to oppose their manipulation by unaccountable institutions whose activities cannot be monitored: a brave but chilling vision, and reinforcing the need to recognise both space and place as conceptual and political issues.

A third attribute of space/place is that each of us is *positioned* or located relative to others. Our positionality depends on who we are – our identity. One aspect of positionality concerns power relationships. We talk about 'knowing your place', which has many possible interpretations. But on the simplest level, positionality implies that there are places where we are welcome and others from which we are excluded by convention or by law because of sex, age, class or colour, or other reasons. Because we all occupy different 'positions', there can be no single, objective account of social situations. 'Learning to see the world from multiple positions – if such an exercise is possible – then becomes a means to better understand how the world as a totality works' (Harvey 1996: 284; see also Haraway 1991).

A fourth type of space is *discursive space*, i.e. a set of mental attitudes and conventions held by members of the public, the media and even, in the example that follows, by the judiciary. An example is that of 'rape space' (Marcus 1992). 'Rape space' is a concept in which women are constructed as potential victims with vulnerable bodies symbolised by the vagina. Such spaces as 'rape space' are discursively produced, i.e. they are mental constructs that are the product of events and the reactions to them, of values and of media debates and representations. The debates that create discursive spaces such as this have the power to influence the behaviour of people, in this case by women avoiding the streets at night.

It is clear that space is not neutral. Far from it! It is actively contested on a daily basis. Space can be contested in many ways: bloodily in wars, with litigation during the planning process, and subtly in the home and work place.

There is a vital link between stages in our development as individuals – our socialisation, or *becoming* (Pred 1984, 1990) – and the locales where this *takes place*. In the office, on the shop floor, in the class room, and at home, we negotiate our self-concepts and identities, not always presenting the same identity in one place as in another, and always evolving in a continuous process of personal development. Thus it is scarcely surprising that Thrift (1983: 28) sees place as an *activity node*; Pollock (1988, quoted by Rose 1993: 112) as the *locus of relationships*; Massey (1994: 120) as *particular moments in . . . intersecting social relations*. Building on Giddens' concept of *locale* as a *setting for interaction*, Thrift argues that such locales 'are the major sites of the processes of *socialization* (seen in the active sense) that take place from birth to death, within which collective modes of behaviour are constantly being negotiated and renegotiated, and rules are learned but also created' (Thrift 1983: 40; see also Giddens 1985).

Among this confusion of place and identity, one concept stands out: that of *home*. Perhaps the most common concept of home is of a material, bounded place where our own activity spaces and those of people closest to us overlap. It is, ideally, where we are most comfortable with our positionality and our relationships with others; a place where we are accepted and affirmed as who we want to be. Home is a discursive space associated with values that overlap – need not necessarily entirely coincide – with the values of those who share it with us. We, and they, often share *collective memories*, that are strongly linked with specific places (Halbwachs 1980). Thus, all the concepts of place suggested so far coincide in the ideal of home. Home is to do with our roots, and is, therefore, a cultural as well as a spatial concept (Hall 1991 – see below). It is a vision that is strongly exclusive of those who do not belong (Sibley 1995). The idea of the nation as home is manipulated by politicians for purposes of developing a sense of unity and mutual commitment. As a concept, home is inseparable from the concept of identity. When our home is destroyed, or irrevocably changed, or is inaccessible to us (after emigration, for example), it can seem as if we ourselves are no longer whole, or are suffering bereavement (Marris 1974: 57). Wendy Chan's chapter in this volume explores what home and the search for it means to Chinese–Canadians from Hong Kong.

This exploration of concepts of space and place brings us finally to yet another abstract, conceptual space: the *chora*. This is a space first conceived by the Greek philosopher, Plato. Its dimensions are far removed from the material or conceptual spaces discussed so far. It exists outside the dimensions of time as measured by the clock, and even of ritual time. Nevertheless, if you are persuaded by Scott Sharpe's argument as presented in his chapter in this volume, it is a space that is as experiential as is material space. In Sharpe's interviews, the experience of being in labour during childbirth can bring with it a sense of the *semiotic* world that precedes the structured, ordered, socio-material, discursive *symbolic* world. This semiotic world is the world of the *chora*, pregnant with as yet unrealised possibilities. In Plato's words, and Kristeva's translation, *chora* is 'a ground for all that can come into being' (Kristeva 1981: 16, footnote 4). Kristeva qualifies *chora* as follows: 'rupture and articulations (rhythm), [it] precedes evidence, verisimilitude, spatiality and temporality' (Kristeva, quoted in Moi 1986: 94). The *chora* exists outside chronological time and outside Euclidean space. However, Lechte gives an additional possible interpretation of the *chora*. Referring to Turner's paintings, he suggests that the *chora* is invoked by the '[s]moke, the light of the sun, ice, steam and clouds . . . ', and argues that in these paintings 'the drive-element of the semiotic gives way to the indeterminacy of topological space' (Lechte, 1995: 101, 102, quoting Serres 1982. See also Grosz 1995). I suggest that one reading of Emily Brontë's poem *Caged Bird*, is to see this poet desperate not just for death but for the liberation of the mind and spirit that is also implied in the concept of the *chora*.

Caged Bird

And like myself, lone, wholly lone,
It sees the day's long sunshine glow;
And like myself it makes its moan
In unexhausted woe.

Give we the hills our equal prayer:
Earth's breezy hills and heaven's blue sea;
We ask for nothing further here
But our own hearts and liberty.

Ah, could my hand unlock its chain,
How gladly would I watch it soar,
And ne'er regret and ne'er complain
To see its shining eyes no more.

But let me think that if to-day
It pines in cold captivity,
To-morrow both shall soar away,
Eternally, entirely Free.
(Emily Brontë, from Roper 1995: 120)

In this section, several concepts of space and place have been examined: the space of (bounded, material) place; activity space, including the space of flows; our positionality in space; discursive space; home; and the idea of *chora*. Access to and use of space varies from person to person, and involves contestation in both public and private arenas. The concept of home brings together all the space/place concepts. Home and the roots that lie there are at the core of our personal identity, even if we are separated from them in time or space. What I have not discussed yet is the body – its space and its use of space.

Geographies of embodiment

> Individual lives are in a sense the smallest possible 'localities', and thus naturally most likely to be situated and constituted in the most contingent and variable ways by combinations of social forces and processes. As a result we tend to dismiss them as largely inaccessible to socio-geographic analyses.
>
> (Hannah 1997: 359)

Why should geographers be interested in the body? And, assuming that some of them are, is it that they are interested in the body for its own sake, or the body in the context of environments? In this section, I will try to answer these questions.

What is the body?

The following definition, quoted in Robyn Longhurst's valuable summary, '(Dis)embodied Geographies' (1997), opens up the complex vistas that geographers encounter by admitting the body into their scope of enquiry:

> By *body* I understand a concrete, material, animate organization of flesh, organs, nerves, muscles, and skeletal structure which are given a unity, cohesiveness, and organization only through their psychical and social inscription as the surface and raw materials of an integrated and cohesive totality . . . The body becomes a *human* body, a body which coincides with the 'shape' and space of a psyche, a body whose epidermic surface bounds a psychical unity, a body which thereby defines the limits of experience and subjectivity, in psychoanalytic terms through the intervention of the (m)other, and ultimately, the Other or Symbolic order (language and rule-governed social order).
>
> (Grosz 1992: 243)

It is clear that the body is far more than mere flesh and bones (even when dead, as Brenda Yeoh shows in this volume). In fact, 'biology with all its imperatives and universals is often only faintly distinguishable beneath the template of symbolic and ritual understandings we lay over it' (Myerhoff 1982: 109). The body – the dynamic assemblage of skin, bones and the rest – is a mere machine without the mind and the *persona* that we develop as a social being under the influence of parents and others. Pithily, Lefebvre (1991: 162) reminds us that 'it is by means of the body that space is perceived, lived – and produced'.

What has the body to do with space?

A simple yet intriguing question! Our bodies occupy space, but they are also spaces in their own right. The 'space' of our body is encoded with 'maps of desire, disgust, pleasure, pain, loathing, love' (Pile 1996: 209). The body and gesture are inseparable: bodies make statements, involuntarily and/or through deliberate choice. The body is a 'site' for consumption and for the expression of values. Through the body's sensory organs, we perceive the qualities of space; through our cultural baggage we assess space; through a combination of creativity and motor skills we adapt and design space. All these statements are undeniable – but they oversimplify. Let me give one example that penetrates far deeper.

We do not always live comfortably with our bodies, and this discomfort may drive us to transform our worlds. The powerful example of fascist men in Nazi Germany is given by Pile (1996), who takes us into the world of psychoanalysis in his description of Theweleit's (1977, 1978) analysis of male Nazi fantasies, of 'men's relationship to the world, growing out of their experiences of their own bodies and their relationship to the bodies of others, located simultaneously in an internalised world and an externalised world' (Pile 1996: 199). Theweleit argues that 'the production of desire and the production of the social are inseparable' and that the particular construction of meaning in the Nazi concept of German nationhood was predicated upon 'a brotherhood of soldiers' united in a shared complex of 'desire, fear, pain and anger', the terrain of which was their bodies and the symbolic and spatial expression of which was the nation in its terrible Nazi form (Pile 1996: 203, 205). Theweleit's ultimate analysis is of 'a body-ego-space under threat of perpetual pain, fragmentation and dissolution' (Pile 1996: 207) in which body was identified with nation – a body-ego-space which, Theweleit warns, remains a perpetual spectre in the postwar world.

Whose body?

An outcome of this book's focus on rites of passage is that several chapters deal with the experiences of those who have commonly been seen as 'Other' – i.e. the object, rather than the subject. Linda McDowell (1992), Vera Chouinard and Ali Grant (1996) and Vera Chouinard (1997) argue that children, the mentally or physically impaired, the unwell, gays and lesbians, abused women, the aged, for example, continue to be overlooked by geographers. In this book, several authors constitute as *subjects* those whose experiences are described and analysed. They provide 'perspectives not as subjugated or disruptive knowledges, but as primary and as constitutive of a different world' (Hartsock 1990: 171). That is, several of our chapters present a specific 'understanding of the world', not claiming to be universal, but established on a foundation of daily experiences, part of the real rather than the 'false *we*' (Hartsock 1990: 171).

Bodyspace, mindspace

Until the 1960s, geography was dominated by the idea that mind and body are separate issues: the mind/body dualism established by the French philosopher Descartes (1596–1650). Reason, society and the state, and culture, were men's domain, and were privileged. The emotions, the body, the family and the individual were assigned as women's sphere, and were secondary issues (Longhurst 1997). Furthermore, Descartes' insistence on isolating rational knowledge from 'the social position of the knower' (Rose 1993: 6–7) was paramount in geography, as in all other social sciences. Not only did the conventional use of 'man' in writings in the social sciences render women and children invisible, but it frequently implied male experience as the norm. Residual categories of the human race were Other (de Beauvoir 1972). So was the body, compared with the mind.

But in the last three to four decades this has changed. The mind/body dichotomy, and the ubiquitous male Subject, have been to an extent swept aside in an invigorating torrent of analysis that has by no means been exclusively from the feminist camps. Pile (1996) traces this back to Kirk (1952) and particularly to Lowenthal's landmark paper of 1961, which established the need for a new epistemology, or theory of knowledge, where geography was concerned, and one that would incorporate experience and imagination. Arguing for putting 'the individual at the centre of geographical concerns' (Pile 1996: 14), and drawing upon psychoanalysis, Lowenthal put forward the idea of profoundly personal worlds as the new *terra incognita* in need of exploration. Shortly afterwards, Kirk (1963) drew on gestalt psychology to develop theoretical concepts that would draw upon subjective experiences, and values, to interpret 'Man and Nature' relationships. Thus began the school of behavioural geography, with its interest in cognitive psychology and how people 'imaged' the world (Boulding 1956).

The meanings that people ascribe to their environments were a major focus of those geographers who went down the phenomenology path. 'Phenomenology provided a people-centred form of knowledge based in human awareness, experience and understanding' (Pile 1996: 50), and geographers such as Buttimer and Seamon recognised that places acquired meaning for people through their activities and personal experiences. Seamon (1980) described how bodies and places were 'choreographed' through regular and repeated patterning of daily routines. The idea that 'human behaviour was founded on shared meanings' (Pile 1996: 53) was central to geographers who worked in the context of yet another school, that of symbolic interactionism. They argued that an individual, relating to selected others in specific places, establishes symbolic patterns through which place is socially constructed. Space, and its significance to the individual, was reaffirmed in this approach. Existentialism, with its deep awareness of the individual's alienation from, or attachment to, place, was another body of thought that began to attract interest from geographers in the 1970s (Samuels 1978).

None of these exciting new approaches came to terms with the intimate issues of the body itself, but, to revert to Grosz's quotation early in this section, they focused either on the psyche, or on the social being. They reflected a 'body of knowledge' that sliced up real bodies for theoretical purposes: the 'outside' was dumped into one set of categories related to corporeality and the 'inside' into another set relating to the mind. Probyn argues that this crude division, 'inside and outside', is better represented by the idea of a pleat. Body and self are intricately folded within each other. Rather than a unity of body and self, there is a doubling: an *embodied self* (Probyn 1991, building upon French philosophers Foucault and Michaux).

Furthermore, the being (the embodied self) that we are discussing here has to be critically thought about ('problematised') by that being itself. And, at the same time, that being is carrying out all sorts of daily routines and responsibilities, i.e. practices. 'The inside and the outside, the ontological[2] and epistemological,[3] the body and the self, sexuality and gender, the practice and the problematization – all are thus intimately articulated through the technologies of the self' (Foucault 1986: 27, quoted in Probyn 1991:21).

If this seems a long way from geographies of the everyday, reading some of the chapters that follow will indicate that 'the search to find/be myself' is a search deeply embedded both in a recognition of what is, and is not, socially acceptable in particular places/contexts (the rules and sanctions of social institutions – see below), what one's body will permit (or what the individual chooses to permit) and the decisions and challenges one feels motivated to follow through. The very action of writing these chapters represents the individual as a reflective self: 'the self is something to write about, a theme or object (subject) of writing activity' (Foucault 1986: 27, quoted in Probyn 1991:21). A significant dilemma that this brings us to is the tension between the uniqueness of one person's experience on the one hand and, on the other, the rights of others to speak for their uniqueness – and that these 'others' are 'subjects' too. Too often, 'other' bodies, 'be they of color, lesbian or homosexual, are made to speak the truth of moral panics' (Probyn 1991: 115).

Thus, the place we arrive at after rejecting the possibility of separating mind from body is that it is not possible to reduce human experience to a limited number of generalisations. Our differences are irreducible, but at least we can see that our knowledge is 'embodied, engendered and embedded in the material context of place and space' (Duncan 1996: 1). The situation is further complicated by the fact that our situated experiences and understandings are in a continuous process of flux throughout the life-course.

The body as a political field and as an agent

We are 'placed' in space but also in the context of social groups and organisations, which we can designate *institutions*. How people are shaped by the discourse of institutions is explored by Foucault in his 'analysis of the practices of exclusion, domination and interiorization of the norms' (Braidotti 1991: 76). The scope of Foucault's work is vast and impossible to summarise here; readers will find Rosi Braidotti's *Patterns of Dissonance* (1991) an appropriate introduction. The main point that needs to be made here is that Foucault defined the body as a *political field*:

> Foucault looks at the institutions, both material (families, schools, prisons etc.) and discursive (the disciplines and other formalized knowledges in the human and social sciences; the penal discourse of the law and so on) that shape the body, the situated, embodied structure of subjectivity.
>
> (Braidotti 1991: 77–78)

The body is both *object* and *target* of power, and the manipulation of the body is carried out in the context of a powerful field of discourse that can be analysed on two levels:

> . . . the procedures designed to subjugate the population, for example the educational, medical and juridical systems.
> . . . the institutions and scientific discourses that ensure control, sanctions and exclusion.
>
> (Braidotti 1991: 80)

However, we are not passively manipulated and subjugated! Anthony Giddens (1984), in proposing structuration theory, conceives of people as *agents*. We act as informed and discriminating agents, but we adopt practices that are constrained by social structures such as *institutions*, e.g. schools, corporations or other employers, the church, the family. Codes of behaviour can also be seen as institutions, as they set behavioural norms; and hegemonic discourses such as the patriarchy are another form of institution. Crucially, Giddens argues that people change those structures through adapting their personal practices. This process of change is a *recursive* process, i.e. turned back on the structures that embody the constraints on behaviour and modifying them. Thus, Giddens recognises the power of social institutions but also the potential of individuals to change them. *Structure* (institutions) and *agency* (individuals) are, thus, mutually dependent. This relationship is referred to as the *duality* of social structures, which are 'both constituted by human practices, and yet at the same time they are the very medium of this constitution' (Thrift 1983: 29). The role of place in this process is:

> a constantly re-energised repository of socially and politically relevant traditions and identity which serves to mediate between the everyday

> lives of individuals on the one hand, and the national and supra-national institutions which constrain and enable those lives, on the other.
>
> (Agnew and Duncan 1989: 7)

Those who have developed structuration theory have been intrigued by the concept of time geography developed by Torsten Hägerstrand.[4] Compiling a time-geography model is a revealing exercise, because the models distil, to an extent, the time–space relationships between agent, place and institution. But, in reducing a 'day in the life of (a person)' to mere patterns in time and space, the time-geography models imply that space is neutral whereas, as pointed out above, it is contested and is by no means equally available and accessible to all. Furthermore, these diagrams cannot represent the activity spaces, the informal networking that makes complex lives possible, and difficult lives bearable (Rose 1993).

Identity

In their valuable introductory chapter to *Mapping the Subject*, Pile and Thrift (1995) introduce five 'territories of the subject': body, self, the person, identity, subject. To conclude this section it is appropriate to delve a little deeper into one of these concepts, that of identity, because the theme of this book, rites of passage, involves examining stages in the becoming of individual identity.

This concept of a dynamic identity, in contrast to 'a monolithic and sedentary image of self and identity' (Pile and Thrift 1995: 10) is associated with such writers as Butler, Castoriadis, Deleuze and Irigaray. Despite – perhaps it is because of – the dynamism of personal identity, Hall (1991) feels that we seek consistency in our identity – seen as self-concept:

> We keep hoping that identities will come our way because the rest of the world is so confusing: everything else is turning, but identities ought to be some stable point of reference which were like that in the past, are now and ever shall be, still points in a turning world.
>
> (Hall 1991: 22)

Hall suggests that this need for a sense of consistent self-concept is coupled with an equally widespread craving for a sense of personal rootedness. Thus, seeking cultural *roots* in a post-modern world becomes a process of following and reconciling several *routes*. Hall (1991, 1995) describes this, for immigrants from the Caribbean to Britain, struggling to establish in the minds of others their simultaneous identity as black, British and with Caribbean ancestry, an identity in which 'otherness, difference and specificity' are blended – sees as a 'postmodernist identity' (Haraway 1991, quoted in Gregory 1994: 162). Identity and culture as well as place are sites of invasion, contestation – and in need of conciliation. Refusing to be 'put in one's place' is characteristic of post-colonial resistance to former dominant paradigms established by colonial powers (Gregory 1994: 163).

King (1991: 14–15), introducing Hall's arguments, sketches the background to the destruction of the taken-for-granted certainties:

> The five great decenterings of modern thought have ended the old logic of identity: Marx, lodging the individual or collective subject always within historical practices; Freud, confronting the self with 'the great continent of the unconscious', making it a 'fragile thing'; Saussure and linguistics pre-empting the process of enunciation; the relativisation of the Western episteme by the rise of other cultures; and finally, the displacement of the masculine gaze. These old collective identities of class, race, nation, of gender and the West no longer provide the codes of identity which they did in the past; existence in the modern world is much more characterised by 'technologies of the self'.

The massive philosophical challenges sketched by King have rocked the edifice of Western philosophy to its foundations. Add to all this the accelerating technological change, and it is not surprising that, just as we seem to crave for home as a secure, familiar and supportive place, so we search also for a sense of stability in our own identity.[5]

'The first and foremost of locations in reality is one's own embodiment' (Braidotti 1994: 161). It is not easy to sum up this section on the body and space. The body is our vehicle for traversing space and for responding to the world's sensory stimuli; it is the location of our psyche, with its drives both creative and destructive; it is the tool we hone in order to communicate, to love and to hate; it offers a 'surface', inscribed by us and read by others; it is a sexed organism that matures, may well become diseased or maimed, and eventually dies; it is a social being on which institutions leave their imprint and by which they in their turn are modified; and which is variously endowed with attributes inherent and acquired (wealth, power and so on). The sort of body that we have prescribes the particular map that we use to navigate our life worlds. Body and self seem impossible to untwine: they are 'pleated' together. Individual identity is by no means fixed, but faces daily challenges to endurance and self-esteem in a world where there are no safe havens, not even in old age, nor in the overarching collective identities of the past. And so this brings us to an examination of the passages that we take and the transitions that we endure on our life's journey.

Rites of passage in the post-modern world

> For groups, as well as for individuals, life itself means to separate and to be reunited, to change form and condition, to die and to be reborn. It is to act and to cease, to wait and rest, and then to begin acting again, but in a different way. And there are always new thresholds to cross: the thresholds of summer and winter, of a season or a year, of a month or a night; the thresholds of birth, adolescence, maturity, and old age; the threshold of death and that of the afterlife – for those who believe in it.
>
> (van Gennep 1960: 189–190)

The classic and influential perspective on rites of passage was developed by Arnold van Gennep in the early years of this century, and each author of the chapters in this book has borne van Gennep's insights in mind as they saw fit. Here, I will introduce the concepts of rites of passage.

In the passage quoted above, van Gennep first refers to biological union, resulting in the creation of a new life, which eventually takes its way independently of its progenitors before itself becoming involved in the cycle of regeneration. He refers then to the diurnal cycle of work and rest; to the solar and lunar cycles; and finally to the life cycle of the individual human body. All this, essentially, is about the structuring of human lives by various natural cycles of time.

Embedded in his vision is the concept of the threshold. The cycles of biological existence are bumpy. Ecologists recognise thresholds that mark the limits of resilience beyond which recovery is possible. Similar thresholds, argues van Gennep, are integral to the life cycles of individual human beings, and, in the societies that he describes, there were culturally appropriate ways of helping individuals win through. These were ceremonial rites of passage, typically – in the pre-industrial societies that he studies – patterned into three stages, which he called a *schéma*. The stages are *separation*, *transition* and *incorporation*. Each of these three stages has its associated rituals. To van Gennep, the *separation* was usually into a sacred world (always a geographical 'zone' or place). Here, those involved existed in a *liminal* state (from the Latin word *limen*, threshold). Their *incorporation* was back into the profane world. He showed that place was an integral part of the rites: 'the passage from one social position to another is identified with a *territorial passage*, such as the entrance into a village or a house, the movement from one room to another, or the crossing of streets and squares' (van Gennep 1960: 192). There was frequently a ritual crossing of a threshold such as passing through a portal of some sort, but this was not necessarily symbolic only, as in everyday life there was often real physical separation of the activity spaces of one group from those of another.

However, the lifecycle thresholds that men, women and children encounter are not all of the same order. Some are connected with biological events (e.g. birth, death). Others mark points of social development (e.g. adolescence/adulthood) or changes in economic role (e.g. retirement). Others are contingent upon unexpected events that can bring about life crises (e.g. accidents, illness, violent attacks, redundancy). Clearly, a 'passage' associated with various types of life crises is still very much part of human experience, but rites associated with individual passages are rare in the post-industrial and post-modern societies that have experienced the 'cultural decentrings' described by King (above).

In materialistic, commercial, wealthy societies, individuals may choose to mark their passage by their consumption patterns, developing new roles by 'acts of disposition and acquisition' (Schouten 1991: 49). Schouten refers to the work of the anthropologist Victor Turner (1974). Without culturally prescribed and shared rites of passage, Turner suggests that people cannot move smoothly from one role to another through the process of separation, transition and incorporation, but linger in a state of liminality. So Schouten argues that one option that these 'liminoids' have is to undertake the construction of new identities alone, using what consumer power and choices they have as a tactic.

Turner offers other concepts that are useful for the theme of this book. He identifies various characteristics of those undergoing the rites connected with the transition stage. Such people are removed for the duration of their liminal stage from the structures of their social group. They represent *communitas*, Buber's Latin term for community, implying thereby a non-hierarchical group of people united by movement towards a goal, but a group that is 'spontaneous, immediate' and unmediated as opposed to the 'norm-governed, institutionalized, abstract nature of social structure' (Turner 1969: 127). Winchester, McGuirk and Everett explore the state of liminality in their chapter on adolescents celebrating the end of their schoolchild status. *Communitas*, argues Turner, is not only associated with liminality but also with marginality and structural inferiority, and all three are conditions which frequently generate 'myths, symbols, rituals, philosophical systems, and works of art' (p. 128) which 'incite men to action as well as to thought' (p. 129). It is in this context that we can interpret Vera Chouinard's descriptions of political grouping and actions of the disabled in Canada in her chapter in this volume.

What we are looking at, then, where rites of passage are concerned, is for the contemporary citizen in a wealthy post-modern society, a process of personal transformation, sometimes revelatory, sometimes agonising, sometimes fun, sometimes requiring a prolonged period of preparation of endurance more like a campaign than a rite of passage, sometimes a lonely, personal experience but sometimes one experienced in the company of another or others. Schouten (1991) argues that, without social support of the kind offered in the past, through the performance of rites, to those undergoing life crises of various kinds, transition stages today can be prolonged, with those involved trapped in a liminoid state.

Our book's journey through the life-course begins with the passage through middle childhood. Margaret Jones and Chris Cunningham examine what is known about play in middle childhood, referring to adult memories, children's literature and a recent Australian case study. Concentrating on 8–12-year-olds, they describe how children continually seek to extend the limits – of the body, of independence from adults, of terrain. Like a bird learning to fly, to handle fickle weather and to elude predators, a child deserves to simply have fun as well as needing to 'spread its wings' through social and physical activity in complex, stimulating environments, including natural ones. There are clear implications, which they spell out, for local government and environmental designers, and a call for research into the world of children's play in the post-modern, hi-tech world.

Adolescent children spend much of their time in school. In the tight confines of an inner metropolitan high school in New York City, Kira Krenichyn shows how teenage boys and girls, limited to the school buildings in their limited time in the lunch-hour, are involved in a gendered conflict over the gym, the only sizeable space available to them for relaxation. Their pent-up physical and psychological frustration is very clear in this account. Teachers' sense of responsibility and difficult choices as gatekeepers of permissible activity also emerge. Interviews show that the powerful expression of physical energy of the boys – demonstrated by their keeping the gym space for their basketball – contrasts with the social constraints imposed on the girls' activities largely because of society's inscription of sexuality on young women's bodies. Excluded from basketball, the girls identify an alternative lunch-hour leisure activity: cheerleading. They are denied this too. Thus, even in the somewhat protected space of a school building, these adolescent girls are already learning the discourses that shape gendered access to space and their limited power to contest those discourses.

These two chapters demonstrate that the concept of play is implicit in the process of socialisation and identity formation. '[P]lay frames allow participants to escape from the "should" or "ought" character of ritual' and permit the fabrication of 'a range of alternative possibilities of behaving, thinking and feeling that is wider than that current or admissible in either the mundane world or the ritual frame' (Turner 1982: 28). Play is, therefore, liberating and potentially subversive.

'Celebrations contain both ritual and play frames', argues Turner (1982: 28). Celebrating the end of school is the topic of the chapter by Hilary Winchester, Pauline McGuirk and Kathryn Everett. They describe how young adults test the limits of their bodily capacities and their capacity for self-control out of the reach of adult relatives, in a setting renowned as a holiday destination for those seeking a good time in the sun, on the surf beaches and in the malls and bars of the Gold Coast, Queensland, Australia. The 'frame' alluded to by Turner, above, refers to the 'enclosure' of ritual or play in 'demarcated times and places', e.g. a temple, theatre or playground (Turner 1982: 28). In this case, school leavers graduate from school room to leisure spot, making it their own for a couple of weeks in the antipodean early summer (November) in a carnivalesque suspension of the rules. Here, in this liminal place at the boundaries of land and sea, where senses are stimulated by limitless horizons of sea, sand and sun, there is a sense of psychological unwinding and blossoming after years of the rigorous time-geography of the school room, a flinging away of clothes and conventions ... thresholds of place and thresholds of identity, a powerful match. Thus, the *communitas* of the about-to-be-inducted exploit their temporary status outside the social structure – neither dependent children nor independent adults – in a brilliant burst of moral licence (see Turner 1969: 109). That the body is the focus, rather than any carnivalesque performances is, arguably, no surprise given the beach context: Lefebvre argues that it is on the beach where the body can behave 'as a *total* body, breaking out of the temporal and spatial shell developed in response to labour' (Lefebvre 1991: 384).

The way the pregnant body is 'mapped' by others is the topic of Robyn Longhurst's chapter. Her study is set in New Zealand. The pregnant body becomes a site for a specific discourse. Instead of being a protected, private space, bodies when pregnant seem, to an extent, to become part of the public sphere. It is the foetus that the women carry that is the focus of this attention, and thus the women find that their own identity is transformed into that of vessel. They are perceived as acquiring irrational, emotional traits, justifying the barrage of 'advice'. Their behaviour is scrutinised in what amounts to an attempt by others to impose limits; and in some jurisdictions they become subject to social sanctions in the form of legislation governing, for example, their smoking or consumption of alcoholic beverages. This amounts to a degree of surveillance which, to a degree, was institutionalised in New Zealand for much of this century through the Plunket Society, which prescribed appropriate forms of mothering behaviour.

Robyn Longhurst's chapter sketches the good-humoured tolerance with which the women deal with the gaze to which they are submitted. For most women, however, this calm nine-month anteroom culminates in one of the most shattering of life's experiences: the birth of a child, '[t]his fundamental challenge to identity' (Kristeva 1981: 31). Medicalised practices and spaces, such as the conventional labour ward, impose constraints on the processes of giving birth – constraints that have met challenge in recent years. Scott Sharpe looks at how Birth Centres, as discursive spaces, offer women more control over birthing. Through interviews with those who have just given birth, he explores the way the body itself 'speaks'. He finds language, and the conventually used conceptual dualisms, hinder the exploration and expression of the birthing experience. Thus, he invokes a pre-discursive, symbolic rather than semiotic world, one which involves cyclical as well as chronological time. He turns to Kristeva's writings on Plato's concept of *chora* and 'lends the *chora* a topology', as Kristeva suggests – in this case, in the Birth Centre.

Parenthood brings with it new rites. It challenges parents to reconsider the nature of their identities and to adapt their time-geographies. Stuart Aitken argues that cultural norms in the USA – and undoubtedly in related societies too – constrain the activities of both mothers and fathers, though more so for mothers. The power relationships are unequal, but, where parenting is concerned, this is not widely acknowledged, as the cultural norms involved are covert. Thus, 'motherhood' and 'fatherhood' are concepts that are deeply embedded in an idealised family structure that has attained the status of a myth. Stuart Aitken argues that 'motherhood' and 'fatherhood' are from 'natural', simple, 'common-sense' concepts. They need to be seen as problematic social constructions.

Maturing into adulthood, giving birth, negotiating the parenthood role, are passages many of us typically go through. The pattern of our lives is no smooth transition, however, but can be ruptured by unpredictable events. Each of the next chapters focuses on one such event.

Rachel Pain has interviewed Scottish women who have been subjected to violent or threatening treatment as children or adults – treatment that has resulted in significant changes in their sense of identity and use of space. It is clear that there are no 'safe places' for women: homes, churches, streets and work places all contain the threat of violence. The picture emerges of women with a shattered sense of security who scrutinise and modify their appearance and their use of space and relationships in order to accommodate this ever-present threat. Unexpectedly, most of these women have retained, or rebuilt, their spatial confidence: 'I won't let it affect me', said one. The caution with which these women regard their options illustrates how the time-geography models referred to earlier are mere scratchings on the surface of social practice. However, harassment in the work place is clearly difficult to challenge. Recent years have seen a greater willingness on the part of women to share their experiences – an unmasking of space.

Space is particularly oppressive to those whose bodies are impaired. The concept of 'ableist space' brings home to us the myriad ways in which those who are not 'fit' are marginalised and excluded. When such people are at the same time women, and/or homosexual, and/or of colour, the picture emerges of a multi-layered oppression, and one for which society through elected governments is less and less willing to take responsibility. As with the women Rachel Pain interviewed, Vera Chouinard responded to her personal experience of impairment by finding ways to challenge her social and institutional context. Looking back on the five years of her own case, Vera Chouinard can now identify the nature of her marginalisation. She was 'out of step' with her colleagues (who assumed she would recover), and with her employers (who were ignorant of their legal responsibilities). She speaks the language of Turner's *communitas* – those outside the social structure – when she comments that the disabled need to adopt as a strategy the *disruption* of spaces of power and privilege. Very similar was the function of the medieval court jester (see Turner 1969: 109–110) and of Shakespeare's fools. The response of power-brokers that 'It's not your place to criticise' is echoed in one of Vera Chouinard's subheadings: 'Being "Out of Place": disabling differences and women's lives'.

Accepting an 'ill identity' is not easy, and involves a difficult process of reconstruction. In fact, the chronically ill have multiple identities, each appropriate for a specific context. Pamela Moss and Isabel Dyck focus on women with M.E. (myalgic encephalomyelitis). Like many other chronic illnesses, 'cultural markers' of M.E. are not always visible to friends and colleagues. The authors interpret the experience of M.E. through the metaphor of the journey – but it is no straightforward journey, and it is not always possible to plan ahead, as episodes of M.E. come and go. They consider the obtaining of a diagnosis – which can take some time – as a rite of passage. Once obtained, the diagnosis can be seen as a script through which the women can begin to make sense of their lives. Pamela Moss and Isabel Dyck propose the concept of a 'radical body politics' as a framework for interpreting the complex experience of illness such as M.E.

Journeys have long been regarded as significant stages of life, rarely, in the past, embarked upon without the performance of specific rites. Passing from one social category, or life stage, to another, was usually accompanied by a change of residence in the societies studied by van Gennep (van Gennep 1960: 192). Migration, of all possible life crises, is a rite of passage in which place, mind/body and identity are bound up in complex ways. Wendy Chan's study of a small group of Chinese-Canadians is primarily a study of their search for home. Those she interviews hold a two-way mirror. One mirror shows the Hong Kong they left. The second shows the Toronto to which they moved, and to which they can return if they choose. Contemporary Hong Kong is now scrutinised in the light of both mirrors. Furthermore, the identity of the person holding the mirrors has been altered by the migrant experience. There is yet another layer to all this, in that most Hong Kong Chinese themselves are descended from emigrants from the Chinese mainland. The nature of their 'Chineseness' is, for these Chinese-Canadians, an issue central to their self-concept and identity, and one they are prepared to discuss frankly. It emerges that, for them, home becomes an issue of identity as much as – even more than – of place.

The next three chapters deal with issues connected with ageing, and the final chapter with death. Richard Hugman explores 'the embodiment of old age as a social construction bounded in space and time' through an examination of three forms of institutional provision for the elderly. The first is residential care, which segregates the elderly and thereby marks them as 'different'. Second, he looks at retirement communities, in which the affluent residents 'consume' a lifestyle which is exclusionary but of which they are in control. Finally, he examines the provision of community care for those who remain in their own homes, another form of dependency but one in which the 'rite of passage' to old age can be managed with a sensitive recognition that 'one's own home' is important to many elderly (where Singapore is concerned, this is not necessarily the old family home, as we will see). Richard Hugman concludes that the elderly body is stigmatised as 'other to' and 'less than' the mainstream of adult life, through an attitude that he labels 'ageism' and which characterises many western societies. The elderly are seen as needing to be separated and managed, their identities 'spoiled' by the ageing process. Yet what is more normal than ageing? He leaves us with the dilemma that 'to untangle ageing and later life from oppression and discrimination is not an easy task . . . '.

What are the effects on the lives of carers when the elderly remain in their own homes? This is the focus of Bonnie Hallman's chapter. She concludes that those who assume responsibility for their parents go through a major life transition. Undertaking eldercare is an unmarked, uncelebrated passage despite the major new commitments that are undertaken. The impact on carers' lives needs to be seen in a gendered context. Although both men and women are deeply involved, women undertake more hours of eldercare and tend to undertake the more personal tasks. With the ageing of the population in countries such as Canada – where this study took place – we need to learn far more about the ways people adjust employment, family life, residence and travel patterns in order to incorporate eldercare responsibilities into their time-geographies.

In Singapore, it is typical for widows and widowers simply to move in with their children. Their old homes, according to Peggy Teo, seem empty, cold and meaningless to them without their spouse. Four-fifths of those over 80 years of age in Singapore live with their children, roughly half regarding themselves as the head of their household, despite the vast majority depending on their children for their income. Here we have the contemporary expression of the institution of Confucian beliefs, which are predicated on strongly hierarchical and patriarchal social relations, requiring the young to respect and care for the aged. From the point of view of the widowed, the Singapore situation is a highly satisfactory accommodation of the widow/erhood identity, in which relinquishment of their old home is compensated for by their participation in the activity space of their family. Here, social and physical thresholds coincide.

Rites to mark the death of an individual undoubtedly have an important function for surviving friends and relatives. In many cultures, however, they are also seen as performing an essential function for the deceased, connected with their passage to the world of the dead, however that is conceived. It is in this latter context that the way the dead body is dealt with assumes great significance. But

Brenda Yeoh introduces a third consideration: the political agenda of the nation-state. Brenda Yeoh investigates various reactions of Chinese-Singaporeans to cremation – strongly promoted by the government since the early 1960s – which now accounts for the disposal of nearly 70 per cent of those who die. The traditional Chinese attitude to death involved choosing a site according to principles of *fengshui*, and, for the surviving relatives, undertaking the age-old practice of sepulchral veneration. The interviewees describe their struggle to reconcile these ancient beliefs with the practice of cremation. Columbaria, where ashes are stored, emerge from this account as far from merely buildings of convenience, but as places that are invested with meanings for the families of those whose remains are stored there – meanings that represent both an inflection and a continuation of traditional Chinese meanings associated with places of the dead.

Some emerging themes

> This is, then, both a social and a very personal landscape . . .
>
> (Probyn 1991: 122)

How can our case studies of embodied geographies enhance and deepen geographical and sociological understandings and imaginations?

First, the chapters that follow make it clear that the transitional stages in life that we have called rites of passage are connected with changes in where those involved are *placed*. Van Gennep titled his first chapter 'The Territorial Passage'; and it was no metaphorical term to him. There is a *territorial passage* for all those whose rites of passage we discuss in this book. It involves learning new socio-spatial patterns that involve time-geography, social positioning, activity spaces and social networks.

Second, learning about place is matched by learning about self. These rites of passage, transitional stages in life, are part of a learning and socialisation process (Litwak 1959) that the individual has to undergo in the course of the development of the self. This is necessary in the course of a life which inevitably involves 'an orderly change from one reference group to another' (Wilson 1980: 140) again and again in the course of new experiences. This learning process comprises the acquisition of new knowledge and practices, many of them spatial. Place and identity are, therefore, a linked part of the learning process involved in the experience of a rite of passage from one life stage to another.

Third, I find it interesting to compare van Gennep's *schéma* (*separation*, *transition* and *incorporation*) with the three stages that Wilson (1980) proposes are gone through by an individual who is adjusting to a new environment. These are, first, the *preparatory stage* of preliminary adjustment of self (which, Wilson suggests, for an immigrant, is that of getting to know local place). The second stage is *the play*, which involves developing socio-spatial networks through place-based social interaction. Lastly is *the game*, in which the individual has developed a consistency in thoughts and behaviour that is maintained despite that individual by then having an extended spatial context (Wilson 1980: 140–144, with reference to several researchers, but especially to Mead 1934). The complete rite of passage would involve moving through the *preparatory stage* into *the play*, and would end when the individual was fully incorporated into *the game*. It is intriguing to broaden the application of this model from 'new environments' to any new life circumstances, such as those encountered in life crises. Then, van Gennep's *separation* can be compared with Wilson's *preparatory stage*; *transition* with *the play*; and *incorporation* with *the game*. In their chapter. Margaret Jones and Chris Cunningham investigate play as a means whereby children, in their *transitional* stage between dependent early childhood and independent young adulthood, develop their mental and motor skills in *preparation* for *incorporation* into *the game* of life as an adolescent and adult. Schoolies Week, a Rabelaisian orgy, would seem to be a ritual of *separation*. Wellington's Birth Exposition, described by Robyn Longhurst, is a ritual of *separation* too, with the pregnant women clearly separated out as targets of a specific discourse. Such women are experiencing the first stage in the complete rite of passage involving labour and childbirth. Birthing is clearly a *transitional* stage, purposeful and requiring team work and mutual understanding between mother and assistants, qualities recognised in the discussion by Wilson of the stage of *the play*. (Scott Sharpe argues that birthing is far more than this – that it is an experience that transcends our routine, day-to-day experiences.) The arrival of the first child in the family, accompanied of course by all sorts of rituals that vary from culture to culture, can be seen as an *incorporation* as far as the new parents are concerned, completing the rite of passage of pregnancy and childbirth. There is possibly a structured approach here to begin to examine the geographies of rites of passage.

Fourth, there may be a latent need for certain *rites* (ceremonies) of passage to be reintroduced. Experiences such as being the victim of violence, 'coming out', disability, or chronic illness involve the individual processes of self-discovery and social adjustment that can be confronting and even traumatic. Solon Kimball, introducing the 1960 edition of *Rites of Passage* comments:

> The critical problems of becoming male and female, of relations within the family, and of passing into old age are directly related to the devices which the society offers the individual to help him [*sic*] achieve the new adjustment. Somehow we seem to have forgotten this – or perhaps the ritual has become so completely individualistic that it is now found for many only in the privacy of the psychoanalyst's couch. The evidence however does not bear out the suggestion. It seems much more likely

> that one dimension of mental illness may arise because an increasing number of individuals are forced to accomplish their transitions alone and with private symbols.
>
> (Kimball 1960: xvii–xviii)

It is clear that the rituals that accompanied the various stages in rites of passage in pre-modern societies were often therapeutic, consolatory, expressions of social support. Compare this with the experience of rape in contemporary society. Could it be that a form of purification rite would heal the psychosomatic wounds that rape involves for women of all cultures? The suggestion that contemporary society could benefit from the reintroduction of rites of passage in certain circumstances is not novel (Crawford 1973; Pollock 1977; Myerhoff 1982; Peters 1994; Pinkola Estes 1997).

A fifth point brings us back to the context of structuration theory. To what extent are those we speak about in the following chapters merely reacting passively to circumstances, in other words as *dupes*? Or to what extent are they creative selves liberated from total reliance on the attitudes of others – a liberation which Wilson suggests is 'one main purpose of existential philosophy' (Wilson 1980: 140)? How do the people featured in our research contribute to the transformation of institutions, including powerful discourses, as they live through the transitional life stages described here? There is plenty of evidence of resistance demonstrated, for example, by pregnant women; women in labour; those women subjected to violence who emerge with strengthened feelings of resistance and confidence; the disabled pressing stubbornly for less oppressive public places for the disabled; those with chronic illness rebuilding their lives; emigrants with a determination to retain their core of ethnic identity; researchers involving those they study in programmes of action research: all these people, all the while, to paraphrase Thrift, are negotiating and renegotiating the social and economic relationships that are the context in which the process of personality unfolds (Thrift 1983: 43).

The sixth and last point relates to 'penetration and the availability of knowledge' – one of the four aspects of 'social action as discourse through and in a region' suggested by Thrift (1983: 42). This point concerns the *re-positioning* of the researched subject as agent. The researcher herself or himself can be an agent in increasing knowledge in certain fields by bringing about a significant learning experience, which can be an integral part of a rite of passage for those co-operating with the researcher.

Thus, rites of passage involve using and learning about new types of spaces and places. We expand the *territorial passage* of van Gennep to include the various types of space/place discussed at the beginning of this chapter. Because our identity evolves through the experiences involved as we encounter new spaces/places, our identity and space/place are inevitably closely linked. Rites of passage in a post-modern world may well be accompanied by a pattern of *geographies of personal discovery* that broadly parallels van Gennep's *schéma* of separation, transition and incorporation. There is a case to be made for considering specific, appropriate rites that might make distressing personal experiences easier to bear. There is evidence that those described in this book are active agents in transforming the institutions that are a part of their lives. The researcher has the opportunity to become part of, and foster, this process. We present, in the following chapters, what Hartsock (1990: 172) calls 'an engaged vision', steps towards the process of 'change and participation in altering power relations'.

Acknowledgements

I should like to thank Dennis Jeans, Jim Walmsley and Bob Haworth, who all read and commented on drafts of this chapter. Of course, what I have finally written reflects my own interpretation of the concepts discussed.

Notes

1 Kimball, introducing the translation of van Gennep's *Rites of Passage*, comments that 'transition' might have been a more appropriate translation of 'passage' (Kimball 1960: vii).

2 'Related to the nature of being'; sometimes referred to as metatheories.

3 'The study of knowledge and the justification of belief' (Dancy 1985, quoted in Johnston *et al.* 1994: 168).

4 Time-geography involves constructing a model of a person's movements through the day, with the two horizontal axes representing, literally, a map of the terrain traversed, and the vertical axes representing time. A unique map of time and space is thus developed that records all the places that an individual visits during the day (Gregory 1985).

5 There is not space in this chapter to introduce and discuss Donna Haraway's exploration of the issue of post-modern identities, and the relations between nature, culture, society and technology, in her book *Simians, Cyborgs and Women: The Reinvention of Nature*. A short gloss on Harraway's arguments is presented in Gregory 1994.

References

Agnew, John A. and Duncan, James S. (1989) 'Introduction', in Agnew, J.A. and Duncan, J.S. (eds) *The Power of Place: Bringing together Geographical and Sociological Imaginations*, Boston: Unwin Hyman, 1–9.

Boulding, K.E. (1956) *The Image: Knowledge in Life and Society*, Ann Arbor: University of Michigan Press.

Braidotti, Rosi (1991) *Patterns of Dissonance*, Cambridge: Polity.

—— (1994) 'Toward a new nomadism: feminist deleuzian tracks; or, metaphysics & metabolism', in C.V. Boundas and D. Olkowski (eds) *Gilles Deleuze and the Theatre of Philosophy*, New York and London: Routledge, 159–186.

Buttimer, Anne (1980) 'Home, reach, and the sense of place', in Anne Buttimer and David Seamon (eds) *The Human Experience of Space and Place*, London: Croom Helm, 166–187.

Castells, Manuel (1989) *The Informational City*, Oxford: Blackwell.

Chouinard, V. (1997) 'Making space for disabling differences: challenging ablist geographies', *Society and Space* 15, 4: 379–390.

Chouinard, Vera and Grant, Ali (1996) 'On being not anywhere near "The Project" ', in Nancy Duncan (ed.), *Bodyspace. Destabilizing Geographies of Gender and Sexuality*, London: Routledge, 170–193.

Crawford, Marion P. (1973) 'Retirement: a rite de passage', *Sociological Review* 21, 3: 447–461.

Dancy, J. (1985) *Introduction to Contemporary Epistemology*, Oxford: Blackwell.

De Beauvoir, S. (1972) *The Second Sex*, Harmondsworth: Penguin.

Duncan, Nancy (1996) '(Re)placings', in Nancy Duncan (ed.) *Bodyspace. Destabilizing Geographies of Gender and Sexuality*, London: Routledge, 1–10.

Foucault, M. (1986) *The Use of Pleasure*, Vol. 2 of *The History of Sexuality*, trans. R. Hurley, New York: Vintage Books.

Giddens, A. (1984) *The Constitution of Society*, Cambridge, Polity.

—— (1985) 'Time, space and regionalisation', in D. Gregory and J. Urry (eds) *Social Relations and Spatial Structures*, Basingstoke: Macmillan, 265–295.

Gregory, D. (1994) *Geographical Imaginations*, Oxford: Blackwell.

—— (1985) 'Suspended animation: the statis of diffusion theory', in D. Gregory and J. Urry (eds) *Social Relations and Spatial Structures*, Basingstoke: Macmillan, 296–336.

Grosz, E. (1992) 'Bodies-cities', in B. Colomina (ed.) *Sexuality and Space*, New York: Princeton Architectural Press, 241–254.

—— (1995) 'Women, *Chora*, Dwelling', in S. Watson and K. Gibson (eds) *Postmodern Cities and Spaces*, Oxford: Blackwell, 47–58.

Halbwachs, Maurice (1980) *Collective Memory*, New York: Harper and Row.

Hall, Stuart (1991) 'Old and new identities, old and new ethnicities', in Anthony D. King (ed.) *Culture, Globalization & the World System*, State University of New York at Binghamton: Department of Art and Art History, 41–68.

—— (1995) 'New cultures for old', in D. Massey and P. Jess (eds) *A Place in the World?*, Milton Keynes: Open University Press, 175–213.

Hannah, Matt (1997) 'Imperfect panopticism: envisioning the construction of normal lives', in G. Benko and U. Strohmayer (eds) *Space & Social Theory*, Oxford: Blackwell, 344–359.

Haraway, Donna (1991) *Simians, Cyborgs, and Women: the Reinvention of Nature*, London: Free Association Books.

Hartsock, Nancy (1990) 'Foucault on power: a theory for women', in L.J. Nicholson (ed.) *Feminism/Postmodernism*, London: Routledge, 157–175.

Harvey, David (1996) *Justice, Nature and the Geography of Difference*, Cambridge, Massachusetts: Blackwell.

Johnston, R.J., Gregory, D. and Smith, D.M. (1994) *Dictionary of Human Geography*, Oxford: Blackwell.

Kearns, Robin (1997) 'Narrative and metaphor in health geographies', *Progress in Human Geography* 21, 2: 269–277.

Kimball, Solon (1960) 'Introduction', in Arnold van Gennep (ed.) *Rites of Passage,* London: Routledge and Kegan Paul, v–xxvi.

King, Anthony D. (1991) 'Introduction: spaces of culture, spaces of knowledge', in Anthony D. King (ed.) *Culture, Globalization and the World System*, State University of New York at Binghamton: Department of Art and Art History, 1–18.

Kirk, W. (1952) 'Historical geography and the concept of the behavioural environment', *Indian Geographical Journal* 25: 152–160.

—— (1963) 'Problems of geography', *Geography* 48: 357–371.

Kristeva, J. (1981) 'Women's time', *Signs* 7, 1: 13–35.

Lechte, J. (1995) '(Not) belonging in postmodern space', in S. Watson and K. Gibson (eds) *Postmodern Cities and Spaces*, Oxford: Blackwell, 99–111.

Lefebvre, Henri (1991) *The Production of Space*, Oxford: Blackwell.

Litwak, Eugene (1959) 'Reference group theory, bureaucratic career, and neighbourhood primary group cohesion', *Sociometry* 22: 72–83.

Longhurst, Robyn (1997) '(Dis)embodied geographies', *Progress in Human Geography* 21, 4: 486–501.

Lowenthal, D. (1961) 'Geography, experience and imagination: towards a geographical epistemology', *Annals of the Association of American Geographers* 51: 241–260.

Marcus, S. (1992) 'Fighting bodies, fighting words: a theory and politics of rape prevention', in J. Butler and J. Scott (eds) *Feminists Theorize the Political*, London: Routledge, 385–403.

Marris, Peter (1974) *Loss and Change*, London: Routledge and Kegan Paul.

Massey, Doreen (1994) *Space, Place and Gender*, Cambridge: Polity.

McDowell, Linda (1992) 'Multiple voices: speaking from inside and outside "The Project"', *Antipode* 24, 1: 56–72.

Mead, George (1934) *Mind, Self, and Society*, Chicago: University of Chicago Press.

Moi, Toril (ed.) (1986) *The Kristeva Reader*, New York: Columbia University Press.

Myerhoff, Barbara (1982) 'Rites of passage: process and paradox' in V. Turner (ed.) *Celebration: Studies in Festivity and Ritual*, Washington DC: Smithsonian Institution Press, 109–135.

Peters, Larry (1994) 'Rites of passage and the borderline syndrome: perspectives in transpersonal anthropology', *Anthropology of Consciousness* 5, 1: 1–15.

Pile, Steve (1996) *The Body and the City. Psychoanalysis, Space and Subjectivity*, London: Routledge.

Pile, Steve and Thrift, Nigel (1995) 'Introduction', in Steve Pile and Nigel Thrift (eds) *Mapping the Subject*, London: Routledge, 1–12.

Pinkola Estes, C. (1997) *Women who Run with Wolves: Myths and Stories about the Wild Woman Archetype*, New York: Ballantine Books.

Pollock, George (1977) 'The mourning process and creative organisational change', *Journal of the American Psychoanalytic Association* 25, 1: 3–34.

Pollock, Griselda (1988) *Vision and Difference; Femininity, Feminism and Histories of Art*, London: Routledge.

Porteous, J. Douglas (1989) *Planned to Death: The Annihilation of a Place Called Howendyke*, Manchester: Manchester University Press.

Pred, Allan (1984) 'Place as historically contingent process: structuration theory and the time-geography of becoming places', *Annals of the Association of American Geographers* 74: 279–297.

—— (1990) *Lost words and lost worlds: modernity and the language of everyday life in nineteenth century Stockholm*, Cambridge: Cambridge University Press.

Probyn, Elspeth (1991) 'This body which is not one: speaking as an embodied self', *Hypatia* 6, 3: 111–124.

Read, P. (1996) *Returning to Nothing*, Cambridge: Cambridge University Press.

Relph, E. (1976) *Place and Placelessness*, London: Pion.

Roper, D. with Chitham, E. (eds) (1995) *The Poems of Emily Brontë*, Clarendon Press: Oxford.

Rose, Gillian (1993) *Feminism and Geography*, Minneapolis: University of Minnesota Press.

Samuels, M.S. (1978) 'Existentialism and human geography', in D. Ley and M.S. Samuels (eds) *Humanistic Geography: Prospects and Problems*, Chicago: Maaroufa Press, 22–40.

Schouten, J.W. (1991) 'Personal rites of passage and the reconstruction of self', *Advances in Consumer Research* 18: 49–51.

Seamon, D. (1980) 'Body-subject, time-space routines, and place-ballets', in A. Buttimer and D. Seamon (eds) *The Human Experience of Place and Space*, London: Croom Helm, 148–165.

Serres, M. (1982) 'Turner translates Carnot', trans. M. Shortland, *Block* 6, 54: 46–55.

Sibley, David (1995) *Geographies of Exclusion*, London and New York: Routledge.

Teather, E.K. (1990) 'Early postwar Sydney: a comparison of its portrayal in fiction and in official documents', *Australian Geographical Studies* 28, 2: 204–223.

—— (1998) 'Voluntary organisations as agents in the becoming of place', *Canadian Geographer* 41, 3: 226–234.

Theweleit, K.(1977) *Male Fantasies*, Vol. 1: *Women, Floods, Bodies, History*, Cambridge: Polity Press.

—— (1978) *Male Fantasies*, Vol. 2: *Male Bodies: Psychoanalysing the White Terror*, Cambridge: Polity Press.

Thrift, N. (1983) 'On the determination of social action in space and time', *Environment and Planning 1: Society & Space* 1: 23–57.

Tuan, Y-F. (1974) *Topophilia: A study of Environmental Perception, Attitudes and Values*, Englewood Cliffs, New Jersey: Prentice-Hall.

Turner, V.W. (1969) *The Ritual Process: Structure and Anti-Structure*, London: Routledge and Kegan Paul.

—— (1974) 'Liminal to liminoid in play, flow and ritual: an essay in comparative symbology', *Rice University Studies* 60, 3: 53–92.

—— (1982) (ed.) *Celebration: Studies in Festivity and Ritual*, Washington DC: Smithsonian Institution Press.

Van Gennep, A. (1960) *The Rites of Passage*, trans. M.B. Vizedom and G.L. Caffee, London: Routledge and Kegan Paul. First published in 1909, *Les Rites de Passage*, Paris: Noury.

Walmsley, D.J. (1988) *Urban Living*, Harlow, UK: Longman Scientific and Technical.

Webber, M.M. (1963) 'Order in diversity: community without propinquity', in L. Wingo (ed.) *Cities and Space: The Future Use of Urban Land*, Baltimore, Maryland: Resources for the Future, Johns Hopkins Press, 23–54.

Wilson, Bobby (1980) 'Social space and symbolic interaction', in Anne Buttimer and David Seamon (eds) *The Human Experience of Space and Place*, London: Croom Helm, 135–147.

2

THE EXPANDING WORLDS OF MIDDLE CHILDHOOD

Margaret Jones and Chris Cunningham (Australia)

Seashells and Sandalwood

My childhood was seashells and sandalwood, windmills
and yachts in the southerly, ploughshares and keels,
fostered by hills and by waves on the breakwater,
sunflowers and ant-orchids, surfboards and wheels,
gulls and green parakeets, sand hills and haystacks, and
brief subtle things that a child does not realise,
horses and porpoises, aloes and clematis –
Do I idealize?
Then – I idealize

(Randolph Stow, 1969)

In this chapter our focus is upon middle childhood – generally between the ages of 8 and 12. We explore the journey through childhood as the child's independent territorial range expands, accompanied by an increasing ability to interpret the physical and social world and to create new worlds in the imagination. While recognising the great differences in childhood experience across the many cultures of the globe, we generally confine discussion to the worlds of children in industrial urban or suburban societies, and particularly Australian society. We look at the role of childhood in human development, and explore issues of play and place; adult recollections of childhood; and the implications of technological and social change on childhood. We question the effects of these changes on the nature of childhood and its needs. Finally we discuss some practical implications of the geography of childhood for urban planning.

The physical experience of the child in modern societies is very different from that of the child in an agrarian, pastoral or hunter gatherer community, and social concepts of childhood are not consistent across time (Aries 1960). In non-industrial societies many childhood activities are soon integrated with the work and play of adult life, initiation into adult society comes at a much earlier age, and the distinction between childhood and adulthood is blurred. The complexity of industrial and post-industrial societies is such that the child will require many more skills to negotiate successfully the diverse facets of living, work and play. Nevertheless, childhood in all societies is, among other things, a time to learn and rehearse physical and interpersonal skills in preparation for entry into adult life.

Childhood has, for past generations of children, been a precarious time. The young were always at greater risk of death or disability from epidemics of disease which swept the world. The industrial revolution brought families to cities in pursuit of work, for adults and children alike. The overcrowded and unsanitary conditions of cities increased the incidence of disease and child mortality.

For children in many parts of the world this is still the case, though the physical well-being of most children in Western industrial societies has, without doubt, improved. Throughout the past century child labour has been virtually abolished, while improved sanitation and medication have decreased the severity of childhood illnesses. Children generally take less part in household work.

There should, therefore, be potential for these children of the more affluent societies of the world to experience opportunities for greater freedom, exploration and self-development than were enjoyed by any previous generation. It is doubtful, however, if circumstances allow them to achieve this potential. Adult priorities intrude on their time and ability to explore their environment freely and independently. The perceived need for education has taken the place of manual work. Free time in childhood is invested, at the direction of well-meaning adults, in learning skills that purportedly contribute towards future material success. There is an underlying feeling of uncertainty about personal safety, especially of children, and this inhibits any dispensation given by adults towards independent action by children (Valentine 1997; Centre of Cultural Risk Research 1998).

While infants are continually exploring and enlarging their domain, it is not until middle childhood that children are able to move, relatively freely, within a wider territory which they can determine for themselves. The challenge is to begin to assert the independence which will lead them to maturity, while still largely conforming to the expectations and limitations defined by their parents, guardians and culture. In this they are constrained, by parental fears and control, by the presence of others such as older children, by community preferences which prohibit those activities enjoyed by children, which interfere with adult priorities, and even by urban planning which promotes adult activities to the detriment of children's mobility. A not-insignificant minority of children is victim to physical, sexual or emotional abuse. The high profile which media reporting and adult concern gives to such abuse further inhibits the freedom which adults are prepared to accord to children generally.

Many experiences of children in so-called advanced societies are common to middle childhood. Children at this time are 'easy'. While they are beginning to join the world at large, this is achieved so quietly that for many children their presence is almost invisible, and their influence on the world is apparently insignificant (Newson and Newson 1986: 142–158). For children themselves, this process is far from insignificant. As their activity space enlarges, their social networks become more complex and their dependence upon adults declines. Middle childhood marks an important passage between dependency and independence. How children's time is used, however, is still largely determined by adults: the daily routine is integrated within the lives of other family members, attendance at school consumes six hours of the waking day, and adult-organised activities such as sport, music, dance, or simply television watching, take up a large proportion of the child's time.

Only the time spent away from these activities – the time to play – seems to belong unequivocally to the child. This time is therefore of particular interest to geographers. While the home remains the centre of the child's universe, the external world, especially as experienced through play, is of increasing importance. As mobility increases, the distinctive territory of the child – the street, the homes of friends, the location of favoured playspaces, places which must be avoided, places which can be easily reached on foot or bicycle, places which are forbidden, places where friends meet, places in which to be alone – gradually develops.

Issues of play and place

Play may be described as the work of childhood (Friedberg 1970; Pearce 1976; Moore 1986a). Through play comes increased self-awareness, realisation of the dynamic nature of relationships with others and the natural world, and an understanding of the position of the child among peers. Social mores are discovered, accepted or discarded. While conflicts with siblings and peers are significant, middle childhood is a relatively peaceful time in comparison with infancy past and adolescence to come. Identity is being developed but not seriously contested, while the relationship between the child and the environment also becomes important.

Children's play may be divided into two groups of activities: the undirected pastimes – 'free play' – which may be undertaken alone or in the company of peers, and recreational activities which are entirely, or to some extent, directed by adults. While adult-directed activities may be enjoyed by the children concerned, and may undoubtedly be socially or educationally useful, the tendency to fill free time this way deprives children of the opportunity to choose activities for themselves.

What do children do when they play? Why do they play? Perhaps the most important reason for free play in middle childhood is the fun and pleasure which it provides. The philosopher, Johan Huizinga (1956), describes play as 'fun', and disputes the *necessity* to relate play to a learning function. The child does not generally see it as primarily a 'useful' activity, although children do seem to understand instinctively how their developmental needs can best be met. Children will play anywhere, but not all play environments provide the same quality of experience. Children especially seek out complex environments where they can interact, alone or with peers, with animals, plants, landforms and artefacts, through exploration, challenge and manipulation of their environment (Hart 1979; Moore 1986b; Nettleton 1987; Cunningham *et al.* 1994).

Games are one of the most important play activities in middle childhood. Throughout recorded history, children's games and activities demonstrate a remarkable similarity, crossing the barriers of both culture and time (Opie and Opie 1969). Immigrant children from distant and differing cultures find common ground with Australian children in playground games (Russell 1986). Children also like 'to keep to the rules', so much so that games such as *Ducks and Drakes*, *Tug of War* and *Blind Man's Buff* have survived for more than two thousand years (Opie and Opie 1969: 6). Children appear to need the constancy of structured games to balance the uncertainty of discovery which embryonic independence provides. Children also choose to impose their own rules in such games, providing a 'gradable challenge' which is within the capability of the child (Eifermann 1971). Evolution and change in these common activities reflect the inventiveness of children.

In play, children experience, and appear to need, a complex mixture of socialisation and physical activity. They also need restful quiet areas where they can contemplate the world in solitude, perhaps to come to terms with change, or simply to withdraw temporarily from their activity-packed lives (Nettleton 1987). Imaginative activities are important. Play areas, particularly away from adult observation, provide the setting for the construction of a personal stage for acting out fantasy situations alone or co-operatively with friends.

Physical activities provide challenge, testing both the individual child's own developing skills and assessing her or his mental and physical ability against those of peers. Informal and organised sport has provided a traditional competitive outlet. Computer games are now popular with children of technologically sophisticated communities and the influence they have on the development of the child will no doubt come increasingly under future scrutiny.

During middle childhood, play relationships are characteristically described as gender specific (Karsten 1998). Boys and girls appear to prefer to play in separate groups, with unflattering epithets directed toward the opposite sex. Nevertheless, in spite of the seeming stigma associated with mixed gender friendships in this age group, mixed gender play is not uncommon, and especially in the street or the wild places which are the natural haunts of children (Opie and Opie 1969; Cunningham and Jones 1997). Differentiation and segregation of children in play may therefore reflect inadequate play environments, or environments used beyond their environmental capacity, rather than an innate propensity.

Adult memories of childhood

Memories of childhood remain clear in the minds of most adults, and those memories may provide ideas for the examination of childhood today. When questioned about favourite places and activities, interaction with the natural world, alone and with peers, is most quickly and frequently recalled (Cunningham and Jones 1994). Edith Cobb (1969: 130) asserts that adult memories of childhood 'refer to a deep desire to renew the ability to perceive as a child and to participate with the whole bodily self in the forms, colours and motions, the sights and sounds of the external world of nature and artefact'. Adult recollections of life as Australian children include fondly remembered childhood sounds and scents of play in natural bushland (as told by biographer, Jacqueline Kent 1988); holiday activities such as exploration, canoe building and playing in the muddy creeks of bushland suburbs (newspaper editor, Hugh Lunn 1989); and author, Ruth Park's, life as a 'forest creature', with the urgent need to explore, the total absorption in games, the existence of her own 'quiet kingdom', and the physical and spiritual influences these had on her (Park 1992).

The universal experience of middle childhood is reflected in children's books, for example, those of Arthur Ransome (1993) and Enid Blyton (1960), which have remained popular with children over several generations, and also in the biographies of their authors. Ransome has based several of his novels on his memories of family holidays at Coniston in the Lake District of the UK, recalling places such as the lake, and boats, rocks and the cowsheds, animals including fish, birds, lizards, moths, caterpillars, newts and farm animals, as well as the charcoal burners, postman, gamekeepers, fishermen and poachers of the village. Vegetation is vividly remembered. Ransome recounts tramping through the hillsides and the importance of secret rituals, situations and adventures which are within the experience of many children today. On the other hand, while Blyton's childhood memories of tending and playing in her large garden are prominent in her biography (Stoney 1974), the Blyton adventure stories are the result of a powerful childhood imagination creating a fantasy world to overcome difficulties experienced in real life. Secret activities and empowered children monopolise the action in her stories, with adults playing supporting roles. Children appreciate the opportunity to be recognised as important members of society, albeit as yet in their fantasy world.

Technological and social change

Are adult recollections of childhood accurate or romanticised? The Australian poet, Randolph Stow, who draws extensively on his own childhood experiences in his writing (Hassell 1990), provides an elegant, if enigmatic, answer. In *Seashells and Sandalwood* (1969: 4), quoted at the head of this chapter, Stow confronts his memories of childhood with the question, 'Do I idealize?' If adult memories of childhood *are* idealised, what is the geography of childhood really like at the end of the twentieth century? Have the rapid technological and social changes of the past decades entirely changed the child's experience of the world?

It is a commonplace that the 'world is getting smaller'. Rapid transport and communication ensure that people are 'closer' to one another than they have ever been. Yet, paradoxically, individuals are, at the same time, more isolated. Community, for adults, no longer means propinquity. Families are scattered, friends no longer live within walking distance, immediate neighbours may be strangers, and children do not necessarily attend schools near home. People travel more and expect greater speed in all that they do. 'Having time', one of the most attractive aspects of childhood, seems to be disappearing in the post-industrial world. It may be argued that efficiency has become the dominant cultural value – to the detriment of play (Godby 1993).

Time which is 'owned' by the child is under threat from the encroachment of adult-driven activities. Cultural, artistic, sporting and social skills are important and recognised as legitimate and worthwhile developmental pursuits, but the child also needs time to confront and address issues, to create and solve problems, to explore the natural world, to experiment with social situations, and time to reflect. The competitive nature of the work environment and the emphasis placed on academic achievement, drive parents to provide their children with supplementary academic education, expensive electronic equipment and educational toys. The warning by a Sydney paediatrician that, 'materialism robs children of imagination, yet parents push it and promote it, forcing children to grow up too fast' (Donaghy 1994: 15), appears not to be heard. Unstructured play apparently has little value in a society where it is expected that material success may be purchased with the currency of formal education.

There can be no doubt that children's independent mobility has diminished throughout the Western world in the second half of the twentieth century. A survey conducted over two decades in England has shown a sharp decline in the mobility of junior school children, in spite of an increase in bicycle ownership (Hillman 1997). A parallel study in Canberra, Australia, has shown a similar and precipitant decline in children's mobility (Tranter 1993). Girls are suffering more restriction within their neighbourhood than boys (Hart 1979; Moore 1986a, 1986b; Matthews 1987, 1992; Cunningham and Jones 1991, 1997; Cunningham *et al.* 1994).

Independent play opportunities, together with development of environmental awareness and social interaction have all suffered because of this declining mobility of children (van der Spek and Noyon 1997). Hillman (1997) found that boys were twice as likely to make the journey to school alone or in the company of another their own age than girls, who were more likely to be accompanied by a parent. While the car has perhaps made it easier for adults to maintain social networks, the quality and frequency of public transport, necessary for the independent mobility of children, is reduced as adult usage declines, and travel costs increase. There is also a perception that public transport is unsafe (Tranter 1993; Cunningham *et al.* 1998). The steady increase in motor traffic has reduced the ability of children to use streets safely. Children are separated from desirable playspaces by major roads, and safe cycleways are rare.

Parental anxieties about children's safety and the changing nature of childhood are the two most significant influences on children's access to independent play (Valentine and McKendrick 1997). This is demonstrated in the diminishing play range of children. Colin Ward, writing in 1977, records instances of children as young as five travelling considerable distances in London independently, and using public transport. A similar story is told by Mary Rose Liverani (1978: 31–32) of travelling with her younger siblings by Glasgow tramways from the inner slums of the city to play in the countryside. It seems, however, that Western children nowadays no longer have the freedom of their city, and that long play ranges are becoming rarer.

A recent study undertaken in Lismore (population 46,000), a major provincial centre in the far north-east of New South Wales, Australia, has thrown further light on the play habits and opportunities of children. Children between the ages of 10 and 12 years, from two schools, were questioned in class about their play activities the previous day. Their parents responded to a questionnaire on their perceptions of their children's after-school play habits, and twenty-four children were provided with disposable cameras to photograph their favourite outdoor play environments and activities. The results were compared with previous data collected from Sydney, New South Wales (4,000,000), Adelaide, South Australia (1,000,000) and Armidale, (22,000), another country town in New South Wales. Table 2.1 shows the strikingly similar play ranges recorded by children in very different urban environments.

Although only 5 per cent of children surveyed in Lismore said their parents considered it unsafe to travel to school independently, children, especially girls, did not range far from their homes to play. Hart (1979) has argued that such gender differences are due both to the additional responsibilities which girls have traditionally undertaken in the home, and also to parental concern for the safety and vulnerability of girls. However, Valentine (1997) has found that parents in the UK are becoming less inclined to expect their daughters to shoulder

Table 2.1 Play ranges of Australian children

Place	*Median play range of girls who played away from home (metres)*	*% of girls who played only at home*	*Median play range of boys who played away from home (metres)*	*% of boys who played only at home*
Armidale	200	31	500	33
Sydney				
Canley Heights	200	85	800	52
Canley Vale	200	71	300	23
Adelaide				
Rose Park	300	77	700	43
Thebarton	100	83	400	76
Semaphore Park	200	63	300	28
Para Hills	250	64	500	58
Para Hills West	150	68	350	50
Lismore	200	71	500	44

Sources: Cunningham and Jones 1991; Cunningham *et al.* 1994, 1996

household responsibilities. The parents concerned also considered girls aged from 10 to 12 to be better able to negotiate public space than boys of similar age because of the greater maturity of the girls and their lesser vulnerability to violence from peers and adolescents. Changing parental perceptions, family composition and community attitudes, including the role of caring for siblings, will, no doubt, continue to modify the spatial opportunities and experience of both boys and girls in future.

Much of the child's life is spent at school, and the schoolyard is one of the most important playspaces for pre-adolescent children. In a sense this is an unnatural play space, accommodating many more children than would freely choose to occupy such a limited area. In ecological terms, its carrying capacity is exceeded. The choice of play activity is largely predetermined by what is provided and what is permitted. Here the conflict between the need for safety and the child's preference for excitement results in a compromise which is very restrictive for children. In the playground, formal and informal sport is popular, especially with boys (Moore 1986b; Jones 1991), providing the opportunity to expend energy and hone physical skills in much the same way that psychomotor skills were practised by girls on fixed metal equipment, that is, before such equipment was removed from parks and playgrounds in the UK and Australia because of concerns about public legal liability (Moore 1986b; Jones 1991). The value of the inclusion of quasi-natural areas for imaginative play, social interaction and contemplation has been well documented (Moore 1986b; van Andel 1986;

Nettleton 1987; Jones 1991). Trees, still valued by children for climbing (Cunningham *et al.* 1996), are being planted again in Australian school playgrounds. This is due to the increasing consciousness of the danger of exposure to the sun, and a consequent awareness of potential legal liability, despite the fact that climbing trees – an almost instinctive activity by children – is forbidden for fear of just such liability.

Rights claimed and exercised by adults can adversely affect those of children. The adult aspiration to own and use the private car may have made children's sporting and cultural extra-curricular activity more possible, but this may also have detrimental effects on the life-style of families, especially of mothers who ferry children from one activity to another (Hillman 1997). Affirmation of the right of both parents to self-expression through their careers has led to increasing professionalisation of child care. Increasing use by parents of long day care and institutional after-school care will certainly influence the geography of middle childhood in the future. It has the potential to restrict opportunities for undirected play and environmental experience. In Australia, the introduction of draft national standards (Commonwealth of Australia 1993) for accreditation of child care facilities has not addressed issues of quality and design of outdoor environments which enhance play opportunities (Australian Play Alliance 1994). Even if generous outdoor playspace standards were applied to such centres, the main issue – that of the ability of children in middle childhood to explore their environment independently – is not addressed. Nor is the answer to the dilemma simple, as it lies in the physical and legal ability of carers to permit children freedom appropriate to their age (taken for granted by earlier generations of children), rather than confining them and tightly organising their play.

Have children and their needs changed? a case study

It is timely, therefore, to ask how the rites of passage of middle childhood have changed over the past generation or two. Are children better or worse off for the apparent curtailment of their freedom consequent upon the advance of post-industrial society with its distinctive mores and physical urban form? Can anything be done to reconcile the apparently different demands on space and time by children and adults? The study by the authors in Lismore was an attempt to respond to these questions. Table 2.2 records the after-school play activities of children surveyed.

Home, generally meaning a detached bungalow on a 'quarter acre' allotment, was shown to be the dominant playspace for Lismore children, particularly girls, after school (Cunningham *et al.* 1996). This finding supports Stretton's (1989: 16) advocacy of this style of housing. The *location* of favourite playspaces in photographs taken by the children also supported the popularity of the suburban

Table 2.2 How Lismore children spent their leisure time on the day prior to survey

Activity[a]	*Girls* n = 47	*Boys* n = 54	*Total* n = 101
Played outdoors at home	26	36	62
Played in street near home	14	16	30
Played outdoors away from home	17	24	41
Played indoors at home	33	27	60
Played indoors away from home	6	9	15
Sport organised by adults	3	4	7
Did homework	24	16	40
Music or cultural pursuits	12	5	17
Watched television at home	29	37	56
Watched television away from home	1	6	7
Played with computer at home	14	19	33
Played with computer away from home	1	1	2
Other	2	6	8

Source: Cunningham *et al.* 1996

Note

a Categories are not mutually exclusive. Children simply recorded the activities they engaged in on the afternoon in question

backyard. In this respect they were no different from the majority of children studied in metropolitan Adelaide, Sydney and country towns elsewhere in Australia (Cunningham 1987; Cunningham and Jones 1996). This 1996 study, however, showed much more similarity between the number of girls and boys who played at home than did the earlier studies. Parents considered the backyard to be the place their children valued most to play, more so than nearby vacant land, parks or the street, although comparison of the photographic evidence (Table 2.3) with results of the parent survey revealed that children played in the street and in open or bushy areas to a greater extent than their parents believed.

Table 2.3 Location of images of play places (percentage of images produced by children)

Location	*% Girls*	*% Boys*	*Total*
Home and backyard	53	54	53
Street near home	6	5	6
Parks	19	14	17
Open or natural areas	22	27	24
Total	100	100	100

Source: Cunningham *et al.* 1996

Table 2.4 Content of children's photographs (number of images produced by children[a])

Setting	*Girls*	*Boys*	*Total*
Home building itself	16	23	39
Formal play area at home (equipment)	43	48	91
Naturalistic areas	76	93	169
Streets and street furniture	7	10	17
Formal play areas	77	25	102

Source: Cunningham *et al.* 1996

Note
a It was possible for a single image to have more than one attribute

No gender bias was shown in the photographs of play locations. Girls and boys largely chose to play in, and photograph, the same sorts of places (Table 2.4).

Both the *location* and *content* of images indicated that natural environments were important to the photographers, though not all children surveyed had easy access to such environments. The choice of *activities* photographed include climbing trees, playing sport, use of play equipment, sitting with friends, playing with animals, riding bikes or skateboards, and playing with toys. Secluded places and cubbies (the Australian children's term for dens or forts) were also photographed, and, on the basis of the children's brief commentary which accompanied the photographs, were valued for their seclusion. Formal sporting facilities featured in many photographs. Mixed gender play was common in the images (Table 2.5).

This study indicated less change in the play *preferences* of children over the past two decades than might be inferred from literature describing the influence of television, computer games and other electronic technology in children's lives. While the impact of private car use and traffic has undoubtedly led to urban forms which restrict the independent play ranges of children, the propensity of children to choose natural play settings where possible is still as strong as Hart

Table 2.5 People in photographs (number of images by sex of photographer)

People in photographs	*Girls*	*Boys*	*Total*
Only boys	8	37	55
Mixed gender	12	14	26
Only girls	28	19	47
Adults present	0	6	6

Source: Cunningham *et al.* 1996

(1979) and Moore (1986b) found with the children they studied. Urban bushland and wild places provided opportunities for solitary or co-operative imaginative and manipulative play. Formal parks, especially those with play equipment which was sufficiently challenging for this age group, were popular. Streets were still used for play, bicycle riding, socialising and 'hanging out'. Informal sport was played in yards, vacant blocks and streets.

Children appear to accept the restriction which adult manipulation of the environment has placed on their freedom, and make the best of what is available to them. At this stage of their life they are unable, or do not have the opportunity, to advocate their own case for the sorts of physical and emotional worlds which best fit their developmental needs. It is therefore imperative that the adults charged with advocating the interests of children – parents, guardians, teachers, urban planners, civic leaders and many others, understand those needs.

Some urban planning implications of taking children seriously

In spite of the propensity for urbanisation there appears to be an increasing acknowledgement of the attachment which exists between people and the natural world. (Kaplan and Kaplan 1989). This is exhibited in such disparate areas as the increasing philosophical interest shown in the environmental movement and the growing commercial tourism industry which promotes wilderness experience and eco-tourism. While such experiences are available to affluent adults, the preservation of natural or quasi-natural areas, forest or bushland, close to residential areas would provide opportunities for children to interact with nature while enhancing the environment for the whole community.

Parks need to serve the interests of all groups, including boisterous and energetic pre-adolescents. However, all too often they also exhibit signs: 'No bikes No ball games No skateboards No dogs', thus proscribing many activities popular with children (Cunningham and Jones 1997). The right of children to engage in such activities must first be recognised, from which the design consequences of such recognition will naturally follow.

The recent trend toward enclosed or 'gated' residential developments reduces access of children to nearby playspaces and opportunities for social interaction. Low fences, Safety Houses and busy street life, on the other hand, can all increase the personal safety and independent mobility of children. Crime reporting by the media in a less sensational way would encourage a more realistic attitude to danger in public streets and spaces for the whole community (Centre of Cultural Risk Research 1998). The street is an important playspace and should be recognised as such in urban design. Independent safe mobility of children is achievable through reduction of traffic speed and volume in residential streets and the provision of footpaths and cycleways which are separate from carriageways. Improvement in public transport systems resulting in increased patronage offers greater safety and mobility for both adults and children. A child-friendly city would therefore be planned with more emphasis on communal space, activity and mobility, and less on individualistic imperatives such as the free right to use motor vehicles.

Despite the demonstrated importance of the home backyard to them, Australian children have few places to meet outside their homes and feel their social needs are ignored. Many traditional community organisations for children, such as Scouts, Guides or church groups, have declined in importance and even ceased to operate in many areas. The need for more and/or better shops for children and a 'kids' night club' was high on the agenda of children questioned in the Lismore study (Cunningham *et al.* 1996). It appears that, among other things, children still seek the sense of community and urbanity that is increasingly being lost from the post-industrial and individualistic city.

Conclusion

Children have no political voice and little opportunity to influence those who plan their environment. Better opportunities for children during the vulnerable passage through middle childhood are not without cost. In an age of so-called economic rationalism the community must examine not only the material costs associated with children, which it frequently does, but also the opportunity costs associated with ignoring the needs of pre-adolescent children as well as the benefits of play to the wider community.

If the passage through childhood is to be a time of joy, discovery and development, then children must be recognised as valuable members of society, in their own right, with needs and contributions uniquely their own, not only as potential adults. Recognition that play is an important element which contributes to the well-being and development of the child, is equally important. Playtime should not be seen as an indulgence, or a reward for 'real' work well done or for good behaviour. For geographers, social scientists and urban planners, the needs of middle childhood could and should influence urban policy and urban form. Unless *children* and *play* command a serious place in the cultural agenda of society, then not only children, but society as a whole will be culturally and spiritually poorer.

References

Aries, P. (1960) *Centuries of Childhood. A Social History of Family Life*, trans. Robert Baldick, New York: Vintage Books.

Australian Play Alliance (ed. C. Cunningham) (1994) 'Outdoor play provision in child care settings', *Twentieth Triennial Conference of the Australian Early Childhood Association*, Perth, Western Australia, 18 September.

Blyton, Enid (1960) *Good Old Secret Seven, the Twelfth Adventure of the Secret Seven Society*, Leicester: Brockhampton Press.

Centre of Cultural Risk Research, Fear of Crime Project Team, Prof. J. Tullock *et al.* (1998) *Fear of Crime*, Australia: Criminology Research Council, National Campaign against Violence and Crime.

Cobb, E. (1969) 'The ecology of imagination in childhood', in P. Shepard and D. McKinley (eds) *The Subversive Science. Essays toward an Ecology of Man*, Boston: Houghton Mifflin, 122–132.

Commonwealth of Australia (1993) *Draft Standards for Centre Long-term Day Care*, Canberra: Department of Community Services.

Cunningham, C.J. (1987) 'The geography of children's play', in R. Le Heron, M. Roche and M. Shepherd (eds) *Geography and Society in a Global Context*, New Zealand Geographical Society Conference Series 14, Palmerston North, New Zealand: Massey University, 245–253.

Cunningham, C.J. and Jones, M.A. (1991) 'Girls and boys come out to play: play, gender and urban planning', *Landscape Australia* 4: 305–311.

—— (1994) 'Child friendly neighbourhoods: a critique of the Australian suburb as an environment for growing up' *Loisir et Société*, 17,1: 83–106.

—— (1996) 'Play through the eyes of children: use of cameras to study after-school leisure time and leisure space by preadolescent children', *Loisir et Société* 19, 2: 341–362.

—— (1997) 'A pitch and a swing: an Australian perspective of urban planning and the child', in R. Camstra (ed.) *Growing Up in a Changing Urban Landscape*, Assen, The Netherlands: Van Gorcum, 119–130.

Cunningham, C.J., Jones, M.A. and Taylor, N. (1994) 'The child-friendly neighbourhood: some questions and tentative answers from Australian research', *International Play Journal* 2: 79–95.

Cunningham, C.J., Jones, M.A., and Barlow, M.K. (1996) *Town Planning and Children: A Case Study of Lismore New South Wales, Australia*, Armidale, Australia: Department of Geography and Planning, University of New England.

Cunningham, C.J., Witherby, A.W. and Cunningham, A.M. (1998) 'Who cares about economic efficiency and quality of life? a discussion of the potential of the most efficient form of urban transport', in *Quality of Life in Cities: Issues and Perspectives*, Vol. 1, Singapore: School of Building and Real Estate, National University of Singapore, 380–395.

Donaghy, B. (1994) Bye bye baby, *Sydney Morning Herald* December 6, 15.

Eifermann, R.R. (1971) 'Social play in childhood', in R.E. Herron and B. Sutton Smith (eds) *Child's Play*, New York: John Wiley and Sons, 270–297.

Friedberg, M.P. (1970) *Play and Interplay*, London: Macmillan.

Godby, G. (1993) 'The future of play', in *World Play Summit: Proceedings*, Melbourne, Australia: International Play Association, 1–22.

Hart, R.A. (1979) *Children's Experience of Place*, New York: Irvington.

Hassall, Anthony J. (ed.) (1990) *Randolph Stow*, Brisbane: University of Queensland Press.

Hillman, M. (1997) 'Children, transport and the quality of urban life', in R. Camstra (ed.) *Growing Up in a Changing Urban Landscape*, Assen, The Netherlands: Van Gorcum, 11–23.

Huizinga, J. (1956) *Homo Ludens*, London: Paladin.

Jones, M.A. (1991) 'Equal opportunity in the playground', unpublished B.A. Honours thesis, Department of Geography and Planning, University of New England, Armidale, NSW, Australia.

Kaplan, R. and Kaplan, S. (1989) *The Experience of Nature*, London: Cambridge University Press.

Karsten, L. (1998) 'Growing up in Amsterdam: differentiation and segregation in children's daily lives', *Urban Studies* 35, 3: 565–581.

Kent, J. (1988) *In the Half Light: Reminiscences of Growing Up in Australia 1900–1970*, Sydney: Angus and Robertson.

Liverani, M.R. (1978) *The Winter Sparrows: Growing up in Scotland and Australia*, Melbourne: Nelson.

Lunn, H. (1989) *Over the Top with Jim*, Brisbane: Queensland University Press.

Matthews, M.H. (1987) 'Gender, home range and environmental cognition', *Transactions of the Institute of British Geographers. New Series*, 12: 43–56.

—— (1992) *Making Sense of Place: Children's Understanding of Large Scale Environments*, London: Harvester Wheatsheaf.

Moore, R.C. (1986a) *Childhood's Domain: Play and Place in Child Development*, London: Croom Helm.

—— (1986b) 'The power of nature orientations in girls and boys toward biotic and abiotic settings in a reconstructed schoolyard', *Children's Environments Quarterly* 3, 3: 52–69.

Nettleton, B. (1987) 'Parks for children: some perspectives on design', *Landscape Australia* 3: 244–248.

Newson J. and Newson, E. (1986) 'Family and sex roles in middle childhood', in D.J. Hargraves and A.M. Colley (eds) *The Psychology of Sex Roles*, London: Harper and Rowe, 142–158.

Opie, I. and Opie, P. (1969) *Children's Games in Street and Playground*, London: Oxford University Press.

Park, Ruth (1992) *A Fence Around the Cuckoo, An Autobiography*, Ringwood, Australia: Viking.

Pearce, J.C. (1976) *Magical Child: Rediscovering Nature's Plan for our Children*, New York: E.P. Dutton.

Ransome, Arthur (1993) *Swallows and Amazons*, London: Red Fox, (first published 1932).

Russell, H. (1986) *Play and Friendships in a Multicultural Playground,* Melbourne: Australian Children's Folklore Publications.

Stoney, B. (1974) *Enid Blyton, A Biography*, London: Hodder and Stoughton.

Stow, R. (1969) 'Seashells and sandalwood', in *A Counterfeit Silence*, Sydney: Angus and Robertson, 4.

Stretton, H. (1989) *Ideas for Australian Cities*, Sydney: Transit Australia Publishing.

Tranter, P. (1993) *Children's Mobility in Canberra: Confinement or Independence?* Canberra: Department of Geography and Oceanography, University College, Australian Defence Force Academy.

Valentine, G. (1997) 'Gender, children and cultures of parenting', in R. Camstra (ed.) *Growing Up in a Changing Urban Landscape*, Assen, The Netherlands: Van Gorcum, 53–80.

Valentine, G. and McKendrick, J. (1997) Children's outdoor play: exploring parental concerns about children's safety and the changing nature of childhood, *Geoforum* 28, 2: 219–235.

Van Andel, J. (1986) 'Physical changes in an elementary schoolyard', *Children's Environments Quarterly* 3, 3: 40–51.

Van der Spek, M. and Noyen, R. (1997) 'Children's freedom of movement in the streets', in R. Camstra (ed.) *Growing Up in a Changing Urban Landscape*, Assen, The Netherlands: Van Gorcum, 24–40.

Ward, C. (1977) *The Child in the City*, London: Architectural Press.

3

MESSAGES ABOUT ADOLESCENT IDENTITY

Coded and contested spaces in a New York City high school

Kira Krenichyn (USA)

The process of identity exploration that occurs during adolescence, the role of physical space in that process, and the ways in which gender becomes intertwined with space as girls, boys and adults attempt to conduct a smooth passage into adulthood, is the focus of this chapter. As Winnicott (1971) observes, identity development throughout the life-course is facilitated by a 'good enough' environment, one which offers a mix of guidance and independence. Such a 'transitional space' allows the individual to play with different identities and ways of being by wilfully acting on surrounding environments, the first of which is the child's home (Aitken and Herman 1997).

The teenagers introduced in this chapter belong to minority groups in the USA. For them, spaces for exploring and examining identity are particularly important, as these teens have grown up in a country that devalues and denigrates its non-white subcultures (Brookins and Robinson 1995; Brookins 1996; Ward 1996; Waters 1996) and that has historically attempted to impose an artificial 'super-identity' upon its many, diverse groups (Erikson 1975). With an absence of formal rites of passage to mark the boundaries, teenagers may become lost in the landscape between childhood and adulthood, and underprivileged minority teens may seek for their missing selves along the easiest avenues, such as gangs, drug trade and early parenthood (Katz 1991; Shakur 1993; Williams and Kornblum 1994; Derezotes 1995; Sikes 1997). Thus, it becomes crucial for minority teens to find opportunities to excel and realise their many potentials, to counter the daily 'challenges to their self-concept' (Brookins 1996: 388), without recycling the very problems from which they hope to escape. This is even truer for girls, most of all girls of colour, who often face a double blow to their self-concept, as they absorb negative messages about both their gender and their race. These girls struggle daily to make sense of those negative messages and envision better possibilities for themselves (Dorney 1995; Benitez 1996; Pastor *et al.* 1996).

While home is the first space where the child locates identity, the adolescent begins to look elsewhere for positive definitions of his or her adult self, to experiment with different possible roles and behaviours. For poor urban teenagers, whose families often cannot finance expensive family trips or recreational activities (Ladd 1978), public institutions like religion, school and recreational organisations may offer alternative opportunities for challenge and self-exploration. Under favourable circumstances, these places support the developmental roles of home and family by helping teens to learn functional skills and forge connections that will help them succeed in the adult world (Derezotes 1995; Jarrett 1995). At the very least, they offer refuge from physically dangerous, hurtful or emotionally restrictive situations found at home or in the outside world (Fine 1995; Taylor *et al.* 1995; Pastor *et al.* 1996).

Of the spaces outside of the home, schools in particular have long been researched for their potential to encourage – or to damage – a student's self-concept, actual achievement and motivation to work toward future goals (see, e.g., Rotheram-Borus *et al.* 1996; Jessup 1967). High school, as one of North America's few rites of passage into adulthood (Fasick 1988), is probably the most available and far-reaching place where teens might reach beyond their own and others' expectations and forge strong adult identities. However, public high schools have also ironically been places where gender, race and class roles are reproduced through inadequate services (Dentler and Elkins 1967; Jessup 1967) and informal practices (see, e.g., Fine 1995), and poor and working class students are socialised to become obedient citizens (Fasick 1988) and 'efficient laborers and citizens' (Aronowitz 1973: 73). In the case of New York City, the government seems to have made a policy of disinvesting in children and schools, which have been notoriously underfunded for years.

Because the traditional school curriculum tends to favour the *status quo*, I would argue that extra-curricular activities offer the best chances for students and teachers to find transitional spaces. The place of sports in high school takes on a particular interest later in this chapter, because sports seem to be among the most challenging and engaging ways to nurture self-esteem, school involvement and ultimate success for a teen (Steitz and Owen 1992; Derezotes 1995; Jarrett 1995). For girls especially, sports and other challenging activities can mean a rare occasion to undo the boundaries of gender identity, to push their physical limitations and to prove that they are capable of accomplishments not usually ascribed to girls (Scraton 1985; Varpalotai 1992; Wearing 1992; Wearing *et al.* 1994; Erkut *et al.* 1996). However, available funding for sports in schools is often directed toward male-oriented activities, male staff and high profile male sports (McRobbie and Garber 1991; Lee 1997), all of which are likely to be intimidating or discouraging for girls. Eder and Parker (1987) put forth a word of caution regarding sports in high schools, pointing out that sports can promote stereotypical gender roles. This reflects an overall tendency for schools to establish gender-specific activities and spaces favouring boys (Thorne 1995; Saegert 1997).

Sports programmes for girls must also be designed with a particular sensitivity to the societal and cultural factors that discourage them from participating in physical activities. These include deep-rooted beliefs about gender-appropriate behaviours and spaces (Hart 1978; Saegert and Hart 1978; Van Vliet 1983; Moore 1986; Matthews 1987), domestic responsibilities (Mauldin and Meeks 1990; Hilton and Haldeman 1991; McRobbie and Garber 1991) and girls' expectations of themselves and their futures (Scraton 1985). Girls may also be restricted by concerns about their physical safety (Katz 1993), especially in a large and chaotic city like New York (Van Staden 1984; Conn and Saegert 1985; Children's Environments Research Group 1992). At puberty new restrictions emerge for girls, as they begin to feel a burden to protect their mature bodies and preserve their sexual reputations (McRobbie and Garber 1991; Cotterell 1993; Taylor *et al*. 1995; DeLeon 1996; Waters 1996) or to cease enjoyable activities and playing with the boys (Stattin and Magnusson 1990; Lee 1997).

Thus, while school has the potential to present a place where poor, urban, minority teens in general, and girls in particular, may resist and overcome the psychological and economic forces aligned against them, mould their future identities, and find their brightest selves, the unfortunate reality is that urban schools have long worked to uphold the same forces that make a viable future so difficult for these teens. Schools leave students and teachers struggling to resolve the difficulties, often with little support. With this conflict as its backdrop, the rest of this chapter explores physical spaces and their corresponding activities within a New York City public high school, and interprets the ways in which boundaries are established around these spaces. While acknowledging that both girls and boys are working to create spaces for identity development, the primary focus is on the interactions among girls, boys and adults as they construct their complicated and varying notions of girlhood, womanhood and the expectations therein.

The study: Neighborhood High, New York City

This chapter is based on research that I conducted as a doctoral student in New York City, with the goal of learning about the different spaces that teenagers use and how these spaces play into the identity development of teens. In order to meet a group of students and teachers who could participate in my research, I volunteered as a teacher's aide in a classroom in a public high school. The school, which I will call Neighborhood High School, is located within a neighbourhood that has long been home to poor and working-class immigrant or minority groups, but a recent real estate boom threatens to displace many of these families. Many of the students live in the neighbourhood, and nearly all are African American or Latino. Those students who do not live near the school travel here from neighbourhoods that are much like this one: run-down, dangerous or drug-ridden.

The school itself, with a total of about 430 students, is small compared to many public schools in New York City. The small size was intentional on the part of the administration, which recently founded the school under a city-wide initiative to nurture progressive education and expand learning opportunities within public schools. Since large and overcrowded classrooms are among the symptoms of New York's ailing public school system, much of the Neighborhood High School's funding is devoted toward teachers' salaries, in order to maximise the number of teaching positions and keep class sizes small. The teachers are young and energetic, and the students are encouraged to address teachers by the first names in order to create a family-like atmosphere and to foster mutual respect. Indeed, one senses a comfortable and safe atmosphere at Neighborhood High, but one of its drawbacks is an absence of special programming, such as music, art, drama or sports, with the exception of a few activities that rely on private donations or volunteerism.

In spite of the efforts to make Neighborhood High unique, it is evident that this school suffers from many of the problems of urban public schools. The large pre-war building shows some signs of neglect, such as broken windows in the stairwells and some of the classrooms. Like most schools of its era, the building's interior is constructed of concrete and ceramic tiles, all of which lend to a drab atmosphere, in spite of attempts to brighten it up with cheerful colours and a few lively murals. The building's oddest feature is the fact that it actually houses two other public schools, one of which is a special school for 'bad' kids. Although I never detected any signs of these other students, their presence seemed to inspire a sense of awe and mystery among the students at Neighborhood High.

Four girls and three boys, 14 and 15 years old, volunteered to participate in focus groups in which we would discuss various activities within the school and their locations. When I met with the girls, they raised many concerns about a lack of opportunities within the school, so they requested that I meet with them again. They also asked me to meet with the school principal, who subsequently offered to meet with the girls. The final meeting with the principal grew from the original focus group but comprised a different mix of girls: two girls from the original group were absent from school, so the remaining two girls attended, one accompanied by her cousin and the other by a friend. Thus, the information that is presented here is based on one focus group with boys, two focus groups with girls, an interview with the school's principal and a final meeting with the principal and four girls, during which I was present. All of these contacts were tape-recorded, transcribed and analysed for common themes that emerged during each session. I have also drawn on field notes that I recorded during my time as a participant observer at the school.

The gym as gendered and contested space

In the focus groups, the school's gymnasium quickly emerged as a meaningful space for many of the students. It was frequently the topic of heated conversations among both boys and girls. The gym was originally a set of adjacent classrooms, which have been renovated by tearing down walls. The result is a space of approximately 30 by 12 metres, divided in halves by a partition with a door to allow passage between the two sides. On each side, basketball hoops have been added at the lengthwise ends, where the windows are located. Large, black metal gates cover the windows behind the basketball hoops, presumably to prevent students from crashing through windows while playing. In the focus group with the boys, they expressed frustration over the design of the space:

JIM: That's not a gym; that's old classrooms!
PEDRO: And then the gym is so small you be catching cramps, you be hurting yourself easy.
JIM: You run into a gate, you fall on to the floor, somebody step on you.
PEDRO: You got gates, you know how many violations we got in this school?
JIM: There, there's ten people a day . . .
PEDRO: You got so many violations, you call the – how do you call that?
KIRA: Inspector?
PEDRO: Inspector – they'll close it down real quick.

The boys also explain that they have little time allotted during the school day for gym class. The only opportunity they have for recreation now is a thirty-minute time slot during their lunch period, when all 430 of the school's students are impossibly expected to play basketball in the cramped gym space. Because of some past infractions committed by students outside school grounds during the lunch period, students are not permitted to leave the building during lunch, and no other spaces or activities are available in the school at this time. Jane, the school's principal, explains that because the building actually houses three separate schools, her students are forced to eat quickly and leave the cafeteria to make room for the next shift of students from one of the other schools. Offering the students a less-than-adequate gym probably seems better to Jane than sending them directly back to class after a short lunch break.

Because of the school's physical constraints, the only students who take advantage of the gym at lunch time are those who are willing to brave the crowded conditions. This excludes most or all of the girls, who end up loitering in the stairwells – sometimes having sex or using drugs there, according to one girl – while Jane single-handedly attempts to clear the staircases in an effort to prevent such activities. This also leaves little room for the boys who are not interested in basketball, like Steven, who prefers to 'just chill [relax]' and 'talk to girls' on a different floor during the lunch period. When asked why he prefers to hang out instead of play basketball, Steven responds, 'why am I just [going to] kill myself and play basketball in that gym? And the odds is I gonna get hurt'. Like the activities in the stairwells, Steven's forays to the other floor are illicit, as Jane has set forth a policy that all students are to be in the gymnasium at this time.

The boys admit that the current circumstances create an unfair arrangement for girls who want to play basketball, but they feel defensive about the small amount of space and time that they are given. As they explain, they are willing to fight for their space, because 'the only chance we get to play basketball we try to take advantage'. They seem to sympathise that some girls would prefer playing basketball over sitting in the stairwells, but the boys understandably want to claim ownership of whatever small space they can acquire. There seems to be a sense of insecurity among the boys' comments, a fear that they might lose what little they have to the girls, who represent a real threat because they have even less.

The low priority toward public school funding, which is unfair to all involved, has resulted in a desperate scramble for a small pool of resources, so that any attempts by teachers or girls to redistribute these resources in a more equitable fashion starts to feel like an injustice to the boys. This sort of struggle is not confined to the gymnasium, as is shown in the following description of an incident that occurred in a classroom:

PEDRO: They [the girls] got a lot down [a lot of favouritism] downstairs with the teachers.

KIRA: They what?

PEDRO: They got a lot down with the teachers. They say something . . . 'cause we got a girl in our class . . .

JIM: They say, they say one thing – boom! The teacher's [yelling at us].

PEDRO: We got a girl in our class . . .

JIM: They take their side and start screaming at the boys (mimics a teacher yelling at the boys).

PEDRO: Right? That specific girl in our class, she whines, she whines and whines.

JIM: (re-enacting the roles of the teacher and the girl) 'I want to sit on the sofa!' 'Get up, let her sit down!'

The type of interplay that is so frustrating for these boys has been noted in at least two other studies (Riddell 1989; Orrn 1993), where girls often attempted to elevate their lower-rung status through exaggerations of typical feminine behaviour, such as flirtation or affectations of helplessness, especially toward male teachers. The boys in this study observe that the girls 'manipulate' the teachers by 'whining', 'crying' and acting like 'babies', and Pedro astutely notes that the girls must sometimes 'fight their way in'. Although they might recognise the underlying dynamics behind these interactions, the outcome is often hurtful and unfair to the boys, whose only offence lies in their collective attempt to find a comfortable space for themselves.

For the girls, injustice has a different meaning. In the gym, the boys fight to hang on to a small space, while the girls struggle to secure a niche within a space already inhabited by the boys. This is a difficult struggle, given that the boys are ready to fight to retain their established territory:

MARIA: The gym is packed.
EVIE: It be packed.
MARIA: You can't even walk through 'cause you're scared a guy's gonna run you over. And if . . . if a little bit get caught they scream at you and they push you out.
EVIE: And they play mad [extremely] rough basketball . . .
MARIA: We try to play volleyball, you know volleyball, right? And we have like two teams, and every time the ball goes to their side they throw it at us [forcefully].
EVIE: And we have to be in the middle because, since they take all the space.
MARIA: There's no room.

The girls stress that they *try* to play volleyball, but that the boys make it nearly impossible, and when they attempt to play basketball the boys steal their ball and throw it at them. They also remark that they would like access to the school's 'game room', a room of about the same size as the gymnasium, which is stocked with art supplies, long work benches and two game tables. The room remains unused during the lunch period, but the girls fear that if they did gain access the boys would probably soon overrun the game room the way they have done in the gym. Instead, the girls continue to search for their own activities and spaces which would be uncontested by the boys, most notably through their recent attempt to start a cheerleading team – which never got off the ground because of Jane's objections. Maria offers an illuminating comment about the embattled nature of the gym space:

> What we wanted to do, was some girls wanted to play basketball. Then they were, you know, they didn't want to. You know they thought it

> was unfair, because they [the boys] thought they [the girls and the teachers] were gonna take some courts from them. But I'm sure they wouldn't – we say cheerleading, they'd be right by our side. There'd be mad [many] boys signing up . . . it's not like we're gonna have room for sports, you know. It's not like we got a big gym and they say you know, we gotta have some sports. Then girls would pick something. But we don't have room . . . we just be like, we don't want the activity. No sports. So we pick cheerleading.

Other girls agree with Maria, commenting that 'It's not that we don't like it [basketball]' and 'At most they [the boys] can play basketball'. To some degree, the girls accommodated their wishes and their expectations of themselves and of the school, once they experienced the backlash that occurred when they displayed an interest in basketball. Thus they looked to cheerleading, perhaps a less preferred activity for some girls, with the hope that they could pursue this interest without opposition.

Cheerleading: 'room to be girls'

Although the girls believed that cheerleading would fit neatly within the boys' physical and psychic space, they were also attracted to the activity itself because it is 'what we like to do'. Due to the volunteer efforts of teachers, the school currently offers two basketball teams, one for girls and one for boys, but some of the girls are not interested in this narrow choice of sports. Cheerleading, which incorporates gymnastics, dance steps and sexy uniforms, might allow them more room to be girls, to have fun, to experience challenge without competition.

Cheerleading may also hold the promise of challenge without the frustration, embarrassment and failure that girls often encounter when they enter into competitive sports for the first time. The girls might be reluctant to play basketball because of past, uncomfortable experiences like the one described by Josie, who was 'ashamed' to have a male gym teacher at her former school. Josie is ordinarily an enthusiastic student, but she often refused to participate in gym class because she simply was not interested in participating in basketball and kickball, the two prescribed activities. She felt that she was being forced to play, and her strained interactions with her teacher escalated to swearing and yelling. Eventually she dropped out of the class. As we were discussing what women and girls might need in order to feel more comfortable, Josie spontaneously stated that 'space' was a key component. Tanya agrees, equating 'space' with 'respect':

> They don't teach us the same way, or they don't treat us with the same kind of respect, you understand what I'm saying? So when there are

> men coaches, they tend to be like – I've noticed this a lot – that they be like, 'Oh, you know, you want to be treated like a man, then you know, then play like a man'. You know what I'm saying, like, they be like, 'All right, you talk about you don't get enough respect, or we don't treat you the same, so if I tell you to do ten push-ups, you doing ten because . . . the guys do it, so you gonna do it, too'. You know what I mean . . .

These girls want the same opportunities as boys, which sometimes means being treated differently from the boys. They imagine a place where they could earn trophies, lift weights, swim, play pool, or just hang out, but without the concomitant gruelling pressures to 'play like a man'. As Tanya clarifies, an activity like cheerleading could provide the perfect arena where girls could excel, carrying their ambitions into academic areas as well:

> You know, sometimes – I'm not trying to say all girls – but sometimes there are girls that don't have nowhere, and either be on the corner hustlin' and smoking weed or drinking, you know have nothing better to do, you understand? Or being with their boyfriend having sex or whatever. Well, now you know, if maybe there was a cheerleading squad a lot of girls would probably be up there, be like 'Oh yeah, I want to make it, you know, I want to make it on the squad, I want to make it on the team, you know I'm gonna try to do good in school'. That would emphasise [motivate] a lot of girls to raise up their grades and try to really look forward to being in school. 'Cause everything is not just the basketball team, 'cause most girls don't like playing basketball, you know?

Marcy, a teacher who had offered to help the girls start the cheerleading team, also feels that cheerleading can be a positive and motivating force in the girls' lives, and she herself was a cheerleader in high school. When the girls began to express an interest in cheerleading, she thought it was a good idea and was willing to coach the team. However, she said, very few girls attended the first meeting, which was surprising to hear, given that the girls had strongly expressed to me that they would like a cheerleading team. When I later asked the girls about the poor turnout, they attributed it to a lack of faith, explaining that they abandoned their efforts 'because nobody believed it could happen, because of the way Jane is; she's like, once she says "no" it's "no"'. Jane's unilateral decision seems particularly unfair to the girls, who believe that the school has failed in its mission to 'make a real school, different from every other school', where students, parents and teachers are all involved in creating a rich and engaging curriculum.

Jane, however, was firmly against a cheerleading team, for one, unwavering reason. She explains her rationale in her meeting with the girls:

> Now, I want to tell you my position about cheerleading, okay? And I'm not going to tell you that it's fair. Okay? But I'll tell you what my feeling is about it, all right? I'm not saying it's the best feeling to have – um, one of my concerns is, you know, the way boys are treated in school, or in life, and the way girls are treated, okay? And one of the things traditionally has been that boys play sports, and girls wear little outfits and go hooray, hooray, hooray. And that they're kind of the bystanders in that, you know what I mean? They're there, but they're not the ones doing it. And I think what happens is sometimes as things go on and on for so many years we just kind of accept it, you know?

Jane emphasises that she does not want the girls to feel that the only legitimate activities for them are those that are secondary to the boys' activities. She wants them to have something that they feel is their own 'because the thing is to be on the team', no matter what team that is. While she would favour activities such as a coed cheerleading, dance or even 'the knitting team', a girls' cheerleading team would reinforce prevalent messages that girls occupy a second-class status.

'Cheerleading . . . is a mind set', Jane says, representing how little progress has actually been made for women in general and for the girls at this school. She feels that boys have found an interest in sports but 'girls have shopping . . . or makeup'. To her, shopping and makeup are problematic, not because they constitute dangerous or lurid expression of the girls' sexuality – she is careful to make clear that she is not 'old-fashioned' or a 'prude' – but because they indicate that so much of the girls' energy is directed toward looking good. Appearance is not just a small feature of a girl's identity, but the only feature, illustrating her desire to attract boys rather than to find her own self. 'I worry', Jane says. 'I think that the majority of girls still see themselves as . . . in terms of what a man will think. And in terms that everything will be all right if they find a man.'

Jane's concerns suggest that many doors have remained closed to girls at this school, and she wishes to help them acquire some of the tools needed to explore the many dimensions of their possible selves, to be 'the ones doing it' and not just the 'bystanders'. Her hopes also encompass all of the students at the school, which she envisions as a more home-like place where students would spend time after school, where they might feel safer and more comfortable. In her interview she discussed her wishes for a greater variety of recreational activities, a safer school facility, elimination of security guards and sofas where students might read, talk or socialise. Overall, Jane conveys a genuine interest in providing comfortable and productive means for girls, and for all of her students, to grow and excel.

However, the girls see her motivations differently, and they have a vague understanding of Jane's objections to cheerleading. They have asked their vice principal, Angel – a Latino man who seems to have gained the students' respect – to clarify the situation, but his explanation fell short of the girls' understanding. Together the girls clamber to grasp the meaning of the adults' actions:

EVIE: But Angel said that's um, sexis– what's it called?
TANYA: 'Sexist'.
JOSIE: Sexual harassment, and . . .
(at this point the girls all begin talking eagerly at once)
EVIE: Because of the skirts.
TANYA: [In disgust] Please!
LANIE [angry]: All right, you want to see a cheerleading skirt? I'm gonna wear one just for her.

The girls' conversation reflects what their experience has taught them about sexism. For them, 'sexism' directly translates into 'sex', and cheerleading is forbidden because of the short skirts, which would mean that others would watch while girls exposed their bodies. Sexism and sexual harassment here mean that girls bear an unfair burden to refrain from doing what they want to do, and the decision against cheerleading seems 'hypocritical' to them, because it means that they must sit still while the boys continue to freely pursue their own activities, free from the worry that they are provoking sexual feelings in others. Their comments throughout the discussion reveal a painful awareness of the double standard:

> If that's the case then boys can't wear tank tops.
>
> So if that's the case, when the boys play ball, they should wear sweat pants and sweat shirts.
>
> Exactly! If the case of so much us being in short skirts, then boys – even though some of them do have busty looking chests – but still . . .
>
> If they look at it that way, well she got a fat bootie, then ya'll don't need to be watching.

Until Jane met with the girls, they knew only one meaning of sexism: restrictions for girls and opportunities for the boys. *Girls* must refrain from doing something they want to do, because *others* (boys, men, families, teachers) might respond inappropriately to their dress and behaviour. Nearly every student, both male and female, has developed a sexually mature body, but only the girls are obliged to conceal the offending parts, the 'busty chests' and the 'booties'. The girls feel that Jane and the vice principal are hypocritical for allowing the boys to play basketball, forbidding the girls to participate in cheerleading, and making the decision without input from the students or their parents. For the girls, this has become another episode of unfairness, lack of choice and restrictions.

Conclusion: social and spatial battle lines

This chapter set out to examine some of the spaces where adolescent development, and girls' development in particular, takes place. In its most basic form, these spaces might offer teens a shelter from the physical dangers of the city streets, and/or they might comprise various places and practices that help them reassemble the pieces of themselves that have been fractured by racism, sexism and other oppressions in their lives (Fine *et al*. 1997). Once this essential ground has been laid, these spaces may continue to expand and multiply, limited only by the girls' and boys' visions of what they can become, perhaps even giving them the tools to chisel away historical boundaries of sex, gender, work and behaviour (Midol and Broyer 1995). A key component of this potential is physical space, in one girl's words, 'somewhere to go during the week' where a teen may realise the 'chance that she might be somebody or do something that she enjoys doing'.

As this story unfolded, it became clear that a small group of teens and adults are fighting to establish a comfortable space within an uncomfortable system, one which provides sparse resources for growth. It was not surprising to find many economic and material restrictions at Neighborhood High, a school which sadly reflects the state of many public schools in New York City. Within this set of limitations, battle lines have been drawn according to gender, in the physical locales of the gymnasium and classrooms, and in the school's social structure. The girls, as a disadvantaged group, contested the boys' claim to territory, by 'whining' and 'manipulating', by appealing to teachers or by attempting to colonise a small area of the gym. When their efforts were unsuccessful, they sought an activity that offered the promise of fun and challenge, along with minimising the chance that the boys would interfere with their pursuits.

The girls were again unsuccessful when they attempted to establish a cheerleading team, only this time they were not hindered by the boys but by Jane. In the absence of an adequate explanation from Jane, the girls drew upon their own knowledge to make sense of the situation, concluding that she would not allow cheerleading because it would be sexually provocative. Given the deeply ingrained nature of taboos around teenage girls and sexual development (Fine 1995), it may be that Jane is anxious about the girls' sexual maturity, unsure how to usher them gracefully into womanhood without encouraging them to be 'bystanders' as adults.

Whatever her inner feelings, the outer expression she has chosen is a protective one. Placed in a difficult position, Jane has chosen to sacrifice the girls' immediate desires, in the hope that she is acting in their best interest. In her meeting with the girls, Jane promised a broader range of activities for the following school year, but she did not confront the difficult reality that the girls often do not receive equal treatment, in school or elsewhere. While she privately voiced concerns to me about the girls' futures, it is not uncommon for someone in Jane's position to publicly deny the obstacles that her students face, afraid she will discourage them if she is too honest (Fine 1995).

It would be unfair to say that Jane has failed her students, as Neighborhood High seems to have succeeded in its goal of providing a safe learning environment, where students greet one another with kisses and chat easily with their teachers. Beneath the weight of a government that withholds money for education, the school has still managed to offer its students some extra-curricular activities, including basketball, self-defence classes for girls and a girls' after-school group.

After Jane met with the girls, they all agreed that Jane's reasoning had some validity. Jane had creative ideas for alternative activities for the girls, such as a dance group, and the girls probably would have been delighted to reach a compromise. With a little work, Jane and the girls might have agreed upon an activity that was challenging and fun, while recognizing that girls do not always subscribe to the rugged competition of many sports.

Most importantly, all of these students – girls and boys alike – have spoken for a need for a space of their own, where they may expand their capabilities and push beyond the expectations that are typically set for these children of the inner city. With few legitimate opportunities at home or in their neighbourhoods, they have turned to school and trusted adults there to help them envision new horizons, but they have been largely disappointed. In spite of its fierce attempts to be otherwise, Neighborhood High School is inescapably a courier of society's voices, and its students are beginning to hear the message that their hopes may be difficult to achieve. Some of the boys – most likely those who are physically stronger – have made basketball their way of succeeding, and they are understandably unwilling to relinquish any of this space to others. In this case, the 'others' are a group of girls, who are learning that they are unwelcome in the gymnasium, or in many of the spheres that have been reserved for males. At this rite of passage, the boys are working hard to secure a physical and public space. The girls, however, are realising that they must locate other spaces as they reach adulthood, and that these will be less visible, less honoured spaces than those occupied by men. How, as women, these girls will construct these spaces, the nature of their contestation, the social structures they will challenge, and the relationship between their spaces and the nature of their women's identities, remains to be seen.

Acknowledgement

I would like to acknowledge all those who have offered their support and insights, especially Susan Saegert, Cindi Katz and Michelle Fine.

References

Aitken, S. and Herman, T. (1997) 'Gender, power, and crib geography: transitional spaces and potential places', *Gender, Place, and Culture* 4, 1: 63–88.

Aronowitz, S. (1973) *False Promises*, New York: McGraw-Hill Book Company.

Benitez, M. (1996) 'Circle of Sistahs', *Reclaiming Children and Youth* 5, 2: 81–86.

Brookins, C. (1996) 'Promoting ethnic identity development in African American youth: the role of rites of passage', *Journal of Black Psychology* 22, 3: 388–417.

Brookins, C. and Robinson, T. (1995) 'Rites-of-passage as resistance to oppression', *Western Journal of Black Studies* 19, 3: 172–180.

Children's Environments Research Group (1992) *Residents' Perceptions of Children's Play Opportunities, West Farms Safe Play Project*, New York: Center for Human Environments.

Conn, M. and Saegert, S. (1985) *Teenagers' Experiences of the Environment at Phipps Plaza South. A Report Prepared for Phipps Houses*, New York: Center for Human Environments.

Cotterell, J. (1993) 'Do macro-level changes in the leisure environment alter leisure constraints on adolescent girls?', *Journal of Environmental Psychology* 13: 125–136.

DeLeon, B. (1996) 'Career development of Hispanic adolescent girls', in B.J. Ross Leadbeater and N. Way (eds) *Urban Girls: Resisting Stereotypes, Creating Identities*, New York: New York University Press, 380–398.

Dentler, R.A. and Elkins, C. (1967) 'Intergroup attitudes, academic performance, and racial composition', in R.A. Dentler, B. Mackler and M.E. Warshauer (eds) *The Urban R's: Race Relations as the Problem in Urban Education*, New York: Frederick A. Praeger, 61–67.

Derezotes, D. (1995) 'Evaluation of the Late Nite Basketball Project', *Child and Adolescent Social Work Journal* 12, 1: 33–50.

Dorney, E. (1995) 'Educating toward resistance: a task for women teaching girls', *Youth and Society* 27, 1: 55–72.

Eder, D. and Parker, S. (1987), 'The cultural production and reproduction of gender: the effect of extracurricular activities on peer-group culture', *Sociology of Education* 60: 200–213.

Erikson, E. (1975) *Life History and the Historical Moment*, New York: W.W. Norton and Company.

Erkut, S., Fields, J.P., Sing, R. and Marx, F. (1996) 'Diversity in girls' experiences: Feeling good about who you are', in B.J. Ross Leadbeater and N. Way (eds) *Urban*

Girls: Resisting Stereotypes, Creating Identities, New York: New York University Press, 53–64.

Fasick, F.A. (1988) 'Patterns of formal education in high school as *rites de passage*', *Adolescence* 23, 90: 457–466.

Fine, M. (1995) *Disruptive Voices: The Possibilities of Feminist Research*, Ann Arbor: The University of Michigan Press.

Fine, M., Weis, L., Centries, C. and Roberts, R. (1997) 'A home of our own: toward a social psychology of spatiality', unpublished manuscript, City University of New York Graduate School and University Center.

Hart, R. (1978) 'Sex differences in the use of outdoor space', in B. Sprung (ed.) *Perspectives on Non-sexist Early Childhood Education*, New York: Teachers College Press, 101–108.

Hilton, J.M. and Haldeman, V. (1991) 'Gender differences in the performance of household tasks by adults and children in single-parent and two-parent, two-earner families', *Journal of Family Issues* 12, 1: 114–130.

Jarrett, R.L. (1995) 'Growing up poor: the family experiences of socially mobile youth in low-income African American neighborhoods', *Journal of Adolescent Research* 10, 1: 111–135.

Jessup, D.K. (1967) 'School integration and minority group achievement', in R.A. Dentler, B. Mackler and M.E. Warshauer (eds) *The urban R's: Race Relations as the Problem in Urban Education*, New York: Frederick A. Praeger, 78–98.

Katz, C. (1991) *A cable to cross a curse: everyday cultural practices of resistance and reproduction among youth in New York City*, unpublished manuscript, City University of New York Graduate School and University Center.

–– (1993) 'Growing girls/closing circles: limits on the space of knowing in rural Sudan and the US cities', in C. Katz and J. Monk (eds) *Full Circles: Geographies of Women over the Life Course*, New York: Routledge, 88–106.

Ladd, F.C. (1978) 'City kids in the absence of legitimate adventure', in R. Kaplan and S. Kaplan (eds) *Humanscapes: Environments for People*, Scituate, Massachusetts: Duxbury Press, 443–447.

Lee, J. (1997) 'Fair game: after 25 years of Title IX, are women really getting an equal crack at the bat? A look at how far we've come and how far we have left to go' *Women's Sports + Fitness* June: 37–40.

Matthews, M.H. (1987) 'Gender, home range, and environmental cognition', *Transactions of the Institute of British Geographers* 12, 1: 43–56.

Mauldin, T. and Meeks, C. (1990) 'Children's time in structured and unstructured activities', *Lifestyles: Family and Economic Issues* 11, 13: 257–281.

McRobbie, A., and Garber, J. (1991) *Feminism and Youth Culture: From Jackie to Just Seventeen*, Cambridge, Massachusetts: Unwin Hyman.

Midol, N. and Broyer, G. (1995) 'Toward an anthropological analysis of new sport cultures: the case of whiz sports in France', *Sociology of Sport Journal* 12: 204–212.

Moore, R. (1986) *Childhood's Domain: Play and Place in Child Development*, Dover, New Hampshire: Croom Helm.

Orhn, E. (1993) 'Gender, influence, and resistance in school', *British Journal of Sociology of Education* 14, 2: 147–158.

Pastor, J., McCormick, J. and Fine, M. (1996) 'Makin' homes: an urban girl thing', in B.J. Ross Leadbeater and N. Way (eds) *Urban Girls: Resisting Stereotypes, Creating Identities*, New York: New York University Press, 15–34.

Riddell, S. (1989) 'Pupils, resistance, and gender codes: a study of classroom encounters', *Gender and Education* 1, 2: 183–197.

Rotheram-Borus, M.J., Dopkins, S., Sabate, N. and Lightfoot, M. (1996) 'Personal and ethnic identity, values, and self-esteem among Black and Latino adolescent girls', in B.J. Ross Leadbeater and N. Way (eds) *Urban Girls: Resisting Stereotypes, Creating Identities*, New York: New York University Press, 35–52.

Saegert, S. (1997) Schools and the ecology of gender, paper presented at the Conference on School Reform, at the Institute for Building and Planning, National Taiwan University, Taipei, Taiwan, May.

Saegert, S. and Hart, R. (1978) 'The development of environmental competence in girls and boys', in M.A. Salter (ed.) *Play: Anthropological Perspectives*, West Point: Leisure Press, 156–176.

Scraton, S. (1985) 'Boys muscle-in where angels fear to tread: girls' subcultures and physical activities, *Sociological Review* Monograph 33: 160–186.

Shakur, S. (1993) *Monster: The Autobiography of an L. A. Gang Member*, New York: Penguin Books.

Sikes, G. (1997) *Eight Ball Chicks: A Year in the Violent World of Girl Gangs*, New York: Anchor Books.

Stattin, H. and Magnusson, D. (1990) *Pubertal Maturation in Female Development*, Hillsdale, New Jersey: Lawrence Erlbaum Associates.

Steitz, J.A. and Owen, T.P. (1992) 'School activities and work: effects on adolescent self-esteem', *Adolescence* 27, 105: 37–50.

Taylor, J., Gilligan, C. and Sullivan, A. (1995) *Between Voice and Silence: Women and Girls, Race and Relationship*, Cambridge, Massachusetts: Harvard University Press.

Thorne, B. (1995) 'Girls and boys together . . . but mostly apart: gender arrangements in elementary school', in M.S. Kimmel and M.A. Messner (eds) *Men's Lives*, Boston: Allyn and Bacon, 61–73.

Van Staden, F. (1984) 'Urban early adolescents, crowding and the neighbourhood experience: a preliminary investigation', *Journal of Environmental Psychology*, 4: 97–118.

Van Vliet, W. (1983) 'Exploring the fourth environment: an examination of the home range of city and suburban teenagers', *Environment and Behavior* 15, 5: 567–588.

Varpalotai, A. (1992) 'A "safe place" for leisure and learning: the Girl Guides of Canada', *Society and Leisure* 15, 1: 115–133.

Ward, J. V. (1996) 'Raising resisters: the role of truth telling in the psychological development of African American girls', in B.J. Ross Leadbeater and N. Way (eds) *Urban Girls: Resisting Stereotypes, Creating Identities*, New York: New York University Press, 85–99.

Waters, M.C. (1996) 'The intersection of gender, race, and ethnicity in identity development of Caribbean American teens', in B.J. Ross Leadbeater and N. Way (eds) *Urban Girls: Resisting Stereotypes, Creating Identities*, New York: New York University Press, 65–84.

Wearing, B.M. (1992) 'Leisure and women's identity in late adolescence: constraints and opportunities', *Society and Leisure* 15, 1: 111–123.

Wearing, B.M., Wearing, S.L. and Kelly, K. (1994) 'Adolescent women, identity and smoking: Leisure experience as resistance', *Sociology of Health and Illness* 16, 5: 626–643.

Williams, T. and Kornblum, W. (1994) *The Uptown Kids: Struggle and Hope in the Projects*, New York: G.P. Putnam's Sons.

Winnicott, D.W. (1971) *Playing and Reality*, New York: Basic Books.

4

SCHOOLIES WEEK AS A RITE OF PASSAGE

A study of celebration and control

Hilary P.M. Winchester, Pauline M. McGuirk and Kathryn Everett (Australia)

Schoolies Week on the Australian Gold Coast is interpreted in this chapter as a rite of passage from youth to adulthood. It occurs in a highly constrained period of space and time, and involves ritualised and transgressive bodily experiences. The spatial context is significant. The Gold Coast is Australia's premier tourist destination, a liminal space separated from the normal workaday environment. The rite of passage of Schoolies Week occurs over several weeks in November–December, the summer holiday period which immediately follows the Australian school-leaving examination, the Higher School Certificate. School students, predominantly from New South Wales (NSW) and Queensland, but also from Victoria, converge on the Gold Coast in a form of pilgrimage to the Australian place which, above all others for Australians, most clearly epitomises sun, surf and sex. We argue here that the liminal location and the temporal separation from school both remove constraints and inhibitions. The result is a week of intense physical activity.

Celebrating the end of school is to the young people involved – the Schoolies – very significant. However, throughout the 1990s, Schoolies Week received negative publicity from the media and from local residents. In 1996, Schoolies Week was increasingly regulated by the Gold Coast City Council (GCCC) under the theme of 'celebration and control', involving both regulation of events and spaces as well as control of adverse publicity. Schoolies Week as an event thus exhibits a tension between liminal behaviour and external control, a tension which has parallels at the individual level.

This chapter draws on four sets of theoretical literature: rites of passage, liminal spaces, Carnival and the body. The theoretical context for this book derives from the concept of rites of passage (van Gennep 1960). In the case of Schoolies Week, the temporal rites of passage occur within, and are inextricably linked to, spaces defined as liminal or 'on the margins' (Shields 1991). The celebratory nature of Schoolies Week also contains elements of Carnival, a concept that implies a huge party which may be an inversion or subversion of normal social relations, but which may also be an 'event' which is an 'authorised transgression' (Bakhtin 1968; Seebohm 1994). Many transgressions of Carnival, as described in the literature, involve bodily experiences, such as cross-dressing (Lewis and Pile 1996). A rite of passage between youth and adulthood necessarily implicates the body; having sex, and getting drunk or stoned to excess, mark a change of state from the imposed discipline of school to the self-chosen freedom to have a body 'out-of-control'.

Liminality and transition

'Rite of passage' was the term first used in 1909 by Arnold van Gennep to refer to those rituals that mark an individual's passage through various stages in the life cycle (Myerhoff 1982: 115). Most of the literature discussing rites of passage calls upon the tripartite structure discerned by van Gennep; separation, transition and incorporation into a new role or status (Myerhoff 1982; Turner and Turner 1982; Schouten 1991). In this case, Schoolies Week can be seen as the transition between school on the one hand and work or university on the other. At a broader level, the transition is between youth and adulthood.

The transition phase is seen as a time of liminality, where identities in the process of transformation become fluid. 'In between the departure from the old position and the incorporation into the new one was a transitional or liminal period, now known as liminality' (Myerhoff 1982: 116). Van Gennep's liminality emphasises the process and the alteration of movements between fixed and fluid in social structure. Shields (1991: 83) claims liminality occurs during 'moments of discontinuity in the social fabric . . . moments of "in-betweeness", of a loss of social coordinates'. 'Classically', Shields also says, 'liminality occurs when people are in transition from one culturally defined stage in the life-cycle to another' (Shields 1991: 83–84).

Liminality is however not just a state and a time, but also has definite spatial connotations. Liminal spaces are those on the margins, whether physical, political or cultural margins (Shields 1991). The beach is an example of a space free from the system of controlling civilisation (Fiske 1989). It has in the past been public rather than private property and has been seen as outside the places of 'rational production' (Shields 1991: 84). Beaches are a type of 'free zone' by virtue of their status as uncertain land, the surface contours of which might change with every tide (Shields 1991: 89). Furthermore, the idea of 'beach' has long been connected to a set of discourses on pleasure and pleasurable activities (Fiske 1989; Shields 1991: 75). The separation of work from holiday space is accompanied by a separation of rationality from desire and of the mundane from the exotic. Within these liminal moments, participants may experience a sense of bonding and community 'found in the mutual relationships of the neophytes in initiation, where community is sacred and serious' (Turner and Turner 1982: 203).

Shields, drawing upon Bakhtin's discussion of the carnivals of the seventeenth and eighteenth centuries, likens liminality to the carnivalesque in its inversion of normative practices wherein all people are reduced to the common denominator of participant (Shields 1991: 89). Carnival may also be considered as 'authorised transgressions' which take place in a controlled and authorised environment (Seebohm 1994: 202–203). As such, carnival can only exist in limited spatial and temporal contexts.

Carnival has been viewed both as a 'ritual of resistance', and as a cathartic 'bread and circuses' safety valve, releasing harmlessly the energies of protest and resistance (Craton 1982). Others see it as (paradoxically) also performing a socially integrative function, distorting the normal relationships between actors and spectators and dissolving the distinction between participant and observer (Bakhtin 1968). Carnival, according to Bakhtin (1968) is not spectacle viewed by people but is lived in by people whom it embraces. Lewis and Pile (1996: 26) argue that carnival is a site of hybridity, independence and ambivalence. Carnival produces a momentary social space based on the politics of pleasure and physical senses (Shields 1991: 95).

The body is the site of transgression, of pleasure and of the senses. Much of the recent literature on the body stresses that the body is a site of resistance and a site of identity (Dorn and Laws 1994; Longhurst 1995). In this case study, the body is used as a vehicle for expression of a changing state. The Schoolies heighten their physical senses by consuming alcohol, drugs and sex. During Schoolies Week, their bodies are not those of disciplined school days, but are undisciplined and 'out of control'.

The Gold Coast is seen here as a liminal space. It lies at the boundaries of land and sea, of Queensland and NSW (Figure 4.1). It is a unique urban environment within Australia, a phenomenon of tourism urbanisation, seen as a city of theme parks and leisure untied to any manufacturing base (Mullins 1993). Its liminality is emphasised by the place imagery and marketing strategies used to define its identity. It is a place to get away. Earlier this century this liminal space was utilised by honeymooning couples who emblazoned their initials on the cliff face at Point Danger, the easternmost headland where NSW and Queensland meet. As a place of liminality, pilgrimage and rites of passage, a parallel may be drawn with Niagara Falls, a place of natural wonder, pilgrimage, honeymooning

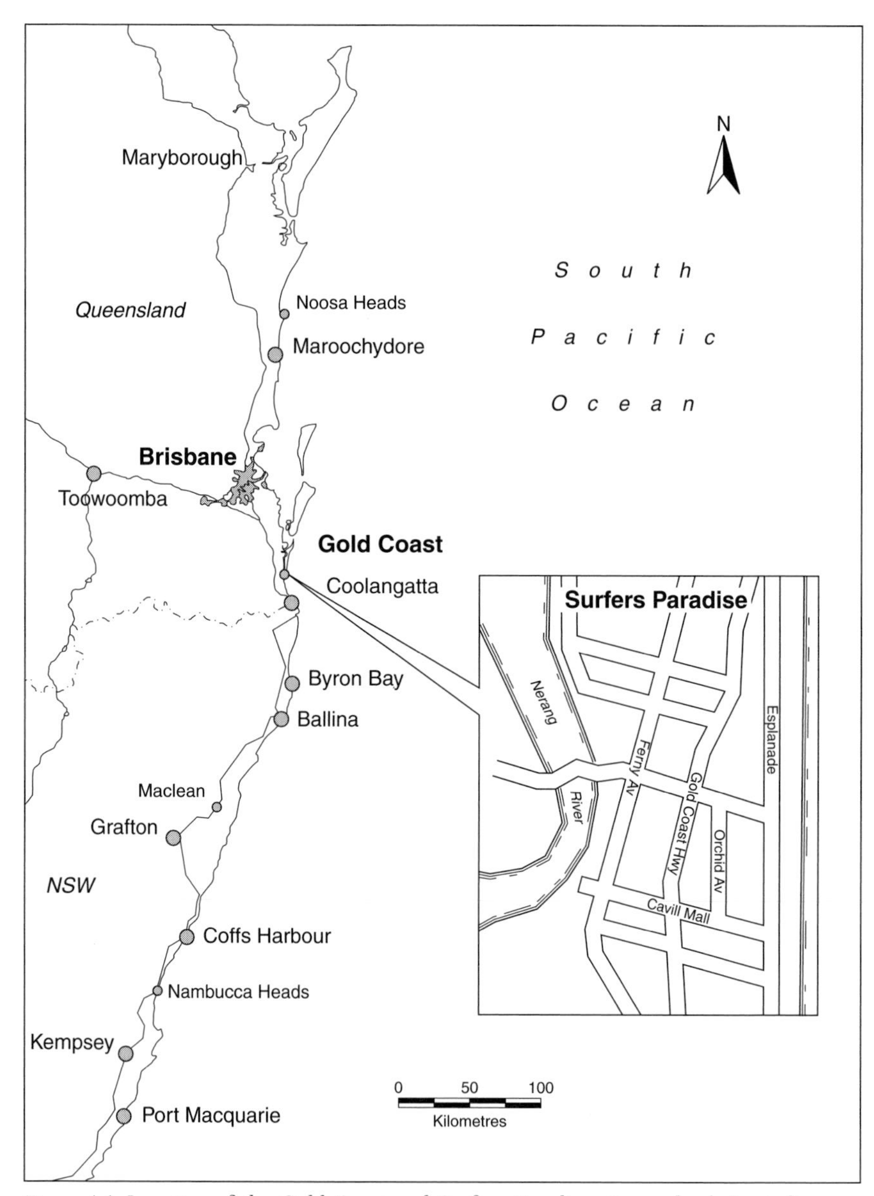

Figure 4.1 Location of the Gold Coast and Surfers Paradise, Queensland, Australia
Drawn by CartoLab, The University of Newcastle

and suicides (Shields 1991). In each place, the natural beauty of the location forms a place of desire and pilgrimage where boundaries are crossed, and identities and states are in transition (e.g. from single to married, from alive to dead). The reputation of the Gold Coast as a location where transitions are possible is what attracts so many Schoolies.

The scale of Schoolies Week is enormous. In the 1990s, approximately 40,000 school students and others have been drawn to the increasingly commercialised event in each of the weeks of late November to early December when, in Australia, the school year ends. The GCCC estimates that it is worth $30–40m to the local economy. For many young Australians, the Gold Coast represents a significant step in the passage to adulthood.

Researching transitions

Researching adolescents' experiences of rites of passage requires in-depth approaches sensitive to the personal nature of how individuals negotiate their transition. The research, undertaken during the first week of December 1996 in Surfers Paradise, the northern suburb of the Gold Coast, employed an intensive immersion technique and a range of qualitative methods. The main methods utilised included participant observation of Schoolies Week and in-depth interviews with participating school leavers. These were supplemented by interviews with controlling agents, i.e. the GCCC and the event coordinator. Some shorter, structured interviews were also undertaken with police and security for perspectives from the controlling agents of hegemonic society and short structured surveys were taken of places of accommodation and local residents. This study also involved the collection and analysis of several forms of textual material relating to the event.

Ethical considerations/power relationships

Four final-year students (three female, one male) were contracted to undertake the field work predominantly for reasons of 'blending in' with the school leavers during participant observation, and also to enable a more peer-like interaction when interviewing the school leavers. Serious consideration was given to the power relations intrinsic in the relationship between researcher and subject as these may strongly influence both access to target groups and the structure and conduct of interviews (Winchester 1996: 122). There is widespread acknowledgement that issues of gender, ethnicity and class, and even of age, dress and language all influence the dynamics of face-to-face interactions (Winchester 1996). For example, in this research the young female researchers found it easier to gain the confidence of the Schoolies they interviewed as they posed no sexual threat. However, these same female researchers occasionally felt patronised when interviewing males in positions of authority.

Participant observation

Participant observation provided the researchers with a minimally obtrusive way of observing Schoolies Week (Spradley 1980; Jorgensen 1989). This method enabled the collection of invaluable information about the social and spatial constructions of Schoolies Week and also enabled the researchers themselves to experience some of the activities in which the school leavers partake. All times of day were observed and participated in. These ranged from the early morning beach scattered with Schoolies left over from the night before, and with early morning risers and other non-Schoolies taking their morning constitutional; through to day-time activities of beaching, pooling and wandering; and to late night/early morning night-club activities and the Cavill Mall scene.

In-depth interviews

There is much advocacy of the rich detail that in-depth interviewing can provide in understanding the meanings and interpretations individuals give to their lives and events within them (see, e.g., Schoenberger 1991; Minichiello *et al.* 1995; Winchester 1999). In-depth interviewing allowed collaborative dialogue between the interviewer and the school leaver, in which new topics could arise freely, with the interviewer mediating the experience and the school leaver holding at least sufficient power to contribute to shaping the content of the discussion without controlling it (Schoenberger 1991). This qualitative method also enabled the school leavers to speak in their own voices rather than using words predetermined by the researcher (Herod 1993: 182).

Thirteen semi-structured in-depth interviews were undertaken with Schoolies. Most were tape-recorded, although noise levels did not always make this possible. These interviewees were approached at random, usually if sitting down waiting or ‘hanging out’ along Cavill Mall, on the promenade or on the beach, and all were undertaken during the day. All were informed of the intentions of the interview and the nature of the research undertaken and informed that they could end the interview whenever they wanted. Semi-structured interviews were also used for obtaining information from the GCCC.

Surveys

In order to supplement information gathered through in-depth interviews, several small surveys were undertaken, providing generalisable data both through forced answer questions and more open-ended questions (McKendrick 1996). Two brief, formal questionnaires directed firstly to managers or front desk staff from twenty-one Surfers Paradise hotel and serviced apartment blocks and secondly to thirty local Surfers Paradise residents. These surveys took the form of directed, formal questionnaires. The two surveys and other informal conversations with security personnel and tourist companies ensured that a broad range of perspectives on the event were brought together, and that the major stakeholders in this event were consulted for their opinions.

Breaking-out

The Schoolies saw Schoolies Week as a celebratory ritual to mark the end of school and a transition from school to adulthood and its associations with work, university and new beginnings:

ADAM: It's sorta like a break between, like next week I'm gonna be starting working, so it's sort of a step from out of school into the workforce sorta thing. It's like a break in between sorta thing.

INTERVIEWER: Yeah ?

KURT: Like, when I finished I'll have no more school left now and it sorta settled in and now I'm gonna have to work. Well this is like the last time we see each other.

(Wollongong Boys)

The trip to the Gold Coast, centring on Surfers Paradise, also took on pilgrimage dimensions as it is marked by the physical separation of these young adults from their parents and adult supervision, from the constraints and inhibitions associated with their home neighbourhoods and the confines of a daily life structured around attending high school. The theoretical significance of a place-based 'letting loose' was recognised by the GCCC's Schoolies event co-ordinator as an important ritual that offers something essential to Schoolies:

> Schoolies is about freedom, it's about having liberty, in some ways, for the first time . . . it is often the first time that the kids are away from home, probably the first time they're on their own holiday, by themselves . . . But all I'm saying is that they're here to celebrate, and it's a very worthwhile and justifiable celebration and it shouldn't be questioned at all, in fact I think it's essential.
>
> (Bruce, event co-ordinator)

For Schoolies themselves, the physical separation in place allowed a loosening of the bonds of their lives as school-goers and a removal from the demands of school and part-time jobs:

ANNIE: Oh, it's more like all your weekends are in, it's like, weekends except you're doing it nine days straight, except we work on weekends so we can't like come to the beach during the day. It's just really good to be away.

FRIEND: It's because you're away from it too, so you're really relaxed. It's not as though you're at home.

ANNIE: Yeah and you're not having to rush off to go to work or you know. You don't have to clean up.

The transition, of which Schoolies Week is a part, is expressed in contradictory and complex ways. Clearly, Schoolies seek to transgress parental authority and the regimented routine of school life through a revelry of intense bodily activity and a relaxed, unstructured celebratory experience: '*a party, to get away from our parents*':

ERIN: Freedom.

DANIELLE: Yeah, no nagging, come home when you want.

ERIN: Yeah, no parents waiting up for you.

DANIELLE: A lot of people had, like a lot of sorts . . . freedom before they came up here, but I didn't. I sorta had my parents down my back all the time watching what I did and everything.

Though a good deal of importance was placed on being able to do whatever they wished whenever they wished, accounts of the daily activities of Schoolies illustrated a semi-routine which might be considered a celebratory inversion of the régime of their school lives: out at night, sleep late, go to the beach, wander and shop on Cavill Mall, perhaps visit a theme park one day, start drinking in their accommodation, go out at night to clubs or to 'spectate' on the crowds on Cavill Avenue. A group of Schoolies from Wollongong gave an outline of a day they considered typical for them:

3: Yeah, well we just get up. By the time that everyone gets up and . . .

1: Showers . . .

2: Has breakfast, then we usually go for a walk uptown.

3: Then go for a walk, maybe get some lunch in town, then um, organise something for tonight, for the night . . .

4: With the other guys as well . . .

3: Then come home, have another shower, get ready, get all spiffed up and then go out.

4: That's it, that's the main sort of thing.

2: Tomorrow we'll go to the beach, spend the whole day at the beach, if it's

> a nice day for it, and then we're going there (points to a theme park brochure).
>
> (Wollongong Boys)

Within this loose structure, freedom of choice is exercised in a rejection of structure which marks the departure from previous routines and the significance of being '*up here by ourselves*':

> We've found that they prefer unstructured events, that they've spent twelve years in a structured environment and prefer to have an unstructured environment when they get here. But if you've got a number of activities on, they will travel around the precinct, they will usually only stay between twenty and thirty minutes in any one particular area. Like they'll go to the dance party for twenty or thirty minutes, get a pass out, go to the rock parties, go down to and watch a bit of the movie and then just meet friends back in the mall there. And like a pinball machine, do it all again.
>
> (Alan Russell, GCCC)

Certainly, Schoolies Week is a phase in which normative practices and performance codes are suspended. Apartment block and hotel managers and security guards relayed stories of lobby parties, discharging fire hoses and extinguishers, broken bottles thrown in swimming pools, balcony hopping, vomiting over top floor balconies, writing obscenities on the walls and trashing rooms. The freedom of anonymity, without the accountability of parental or school supervision, allows behavioural limits to be stretched if not erased:

MARTY: Oh, nobody knows us . . . like, it's totally different up here, you can just go, like really sick, you know. Like, you go into town on a Saturday night and you think, you think, 'there's an old score over there', and you think, 'no, I'm not gonna go over there and make an idiot of myself', but up here you just think, 'nah, I'll just go sick', you know, 'if she bars me I'll just laugh because I'll never see her again'.

INTERVIEWER: So you just feel free?

MARTY: Yeah, don't care, just run amuck. Choice!

Likewise, one group of Schoolies, encountered late one night ripping out signposts and trees, explained that they were doing it 'because we're allowed to trash the place'. The anonymity of the separation from the bounds of home and childhood temporarily changes behavioural norms. Schoolies' identities, normally performed within structured codes and practices, are temporarily unleashed to become fluid during this ritually marked passage.

Embodied experience

The Schoolies phenomenon is highly focused on intense embodied experiences in which the physical senses are stimulated through an inversion of the normatively controlled body to become the out-of-control body. The temporary fluidity of identities in the process of transformation allows experimentation with behaviours considered inappropriate to the cultural definition of school children. For many Schoolies the activities they undertake are similar to those at home – if considerably less restrained. Nevertheless, the intensity of bodily sensation supports the notion of a libidinal/rational separation which differentiates the Schoolies Week from the everyday (Shields 1991).

Despite the fact that a majority of Schoolies are under the legal age to consume alcohol, excessive alcohol consumption is a major part of the ritual representing how the 'normal' rules are suspended during this time. The Wollongong boys considered the ritual of drinking to be a major part of Schoolies week:

1: Alcohol, yeah, it was, it is, yeah. All the other people, all alcohol, alcohol. You go down to the clubs, um the bottle-o down here, and you just see everyone in there, yeah, people carrying their beers. People carrying bourbon.
2: People just wanna get smashed . . .

(Wollongong Boys)

Alcohol without a doubt is the most popular intoxicant for the Schoolies. It lowers inhibitions for most of the socially self-conscious teenagers, allowing them to bluff their way into night-clubs or approach others to flirt or 'pick-up'. It was readily available and easily accessible. Those who were under 18 generally had older friends who could obtain alcohol for them, or used fake identity cards to gain entry to clubs where alcohol was sold. For the many under-18-year-olds denied access to the clubs on Cavill Mall and Orchid Avenue, drinking in their accommodation or outside heavily policed areas was common.

Interviewees indicated that drugs were available; some had seen them, some had not, some were using, some were not. The heavy police presence in public areas was thought to make 'scoring' problematic, while the highly publicised death of one young person and critical illness of several others due to ingesting Fantasy at a Gold Coast night-club just weeks before the interviews took place, may have induced some wariness among Schoolies. One interviewee responded to the mention of drug use with: 'Fuck that shit. I've had a few cones and that, but none of that poppin' e's and shit, fuck that' (informal night interview).

Sex, commonly associated with the 'sun and surf' of the Gold Coast (Jones 1986) is said to be another physical stimulation sought by school leavers during Schoolies Week. In recognition of this, the Health Department had, in previous years, issued Schoolies with 'Willie Wallets' containing condoms as a promotion of safe sex. In 1996 Health Department officials claimed to have visited every Year 12 Student in Queensland providing 'Home Safely' kits which included information on safe sex.

Certainly, flirtation and sexual propositioning is an accepted part of the social theatre of Schoolies: 'oh man, you can't walk down the street without someone either trying to pick you up or you know, something like that' (Anne and friend). Some female Schoolies argued that many of the male Schoolies were 'sleazy' and 'perverts'. The desire to 'pick-up' was clearly part of the agenda of many Schoolies and the various friends who accompany them. As one Schoolie put it: 'One of our friends in the other rooms said it was like a big rooting week . . . She says that's all she comes up here for, you know' (Danielle). Two male Schoolies argued that they were disappointed by the high male to female ratio. For others, 'picking-up' appeared to be less important:

JOEL: Oh, I don't know, I find I have come for a holiday but if you do get it, it's a bit of a bonus.
INTERVIEWER: So you don't go out looking for it?
JOEL: Oh yeah, but it's not the sole purpose.

Nevertheless, many expected at least the possibility of a sexual encounter during their celebratory week:

ANNE: Cause I've got a boyfriend at home, I'm not supposed to be picking up while I'm up here, but I thought, you know, you're gonna be drunk . . .
FRIEND: If it happens, it happens.

There is however some ambivalence and tension exhibited within this temporary excess of freedom from the constraints of normative practice. Responsibility for the outcomes of the explosion of bodily sensation rests, perhaps for the first time, with Schoolies themselves. The balance between the bodily revelry of the week and the burden of this responsibility creates an interesting tension which, in one sense, is a tension between fixity and fluidity. This may account for the manner in which Schoolies generate internal controls in which they adopt informal routines in their daily movements and a palpable concern for their own safety and well-being even during this transitional phase of an excess of 'liberty'. As one Schoolie put it: 'It's a hell of a responsibility being so far away from home. I don't know, its a lesson in life. There are things you do that you think "ho I'd never do that again" ' (Rebecca). While another expressed a similar sense in: 'I feel like, sort of gotta . . . fulfilled or something . . . For myself, I feel just being by yourself, you know. You gotta take care of yourself' (Wollongong Boys).

However, for some, any sense of reaching maturity was overwhelmed by the desire to transgress authority and resist the trappings of maturity:

INTERVIEWER: Do you feel you are in any way maturing this week?
KURT: I'd say immatured (laughs).
INTERVIEWER Well, if you've immatured, what are you doing that you don't normally do?
KURT: Drinking a lot more.
ADAM: Acting stupid.
KURT: . . . we just don't care any more.

The sense of responsibility for personal well-being is a key element in any transition to maturity being undergone by Schoolies. If the week represents a release from the external controls of parental and school supervision, there is evidence of a replacement of this control by the exercise of considerable internal control, exerted to ensure personal and group safety. Rojek (1985) noted that people participating in carnival are openly interdependent upon each other for their bodily sustenance, recreation, pleasure, safety and well-being. This interdependence is evident in a clear awareness among Schoolies of the need to stay aware, keep track of each others' whereabouts and avoid potentially troublesome situations:

INTERVIEWER: So you are aware of your actions?
REBECCA: Yeah you really look and see cops yelling at other people and you think 'go' and you walk the other way. You're not overdoing the goody goody thing but you don't do anything stupid.
KIRSTY: I mean we're pretty responsible, we have a group and we go out and come home together . . . we don't split up or anything.
ANNE: . . . and then we went, 'oh where'd she go . . .' and then we spent our night looking around to see if she's alright. We didn't know if she had a key or anything like that, but she eventually came back.

This sense of responsibility for group well-being was especially evident regarding the 'hangers-on'– usually older males there to 'harry-it'. These 'predators' (Alan Russell, GCCC) were widely acknowledged as a source of the majority of trouble during Schoolies:

ANNE: The Schoolies ones are alright, but it's like the 24-year-olds and that . . .

FRIEND: Yeah that's right. We always ask people if they're Schoolies people because we'll be nice to them. But all the guys who are up here, just because they know Schoolies girls are up here, they're real dickheads. You know, they're up here for one thing and they're all being arseholes and that, and they're rude.

This behaviour lies in tension with the sometimes reckless abandon with which Schoolies conduct themselves during their liminal celebration. Certainly, there was evidence of Schoolies forming bonds with other Schoolies to ensure personal safety and unifying against non-Schoolies who were perceived as outside their ritual and therefore potentially troublesome: 'It's like a code of honour . . . (a) hidden curriculum of communication here during Schoolies' (Bruce, event co-ordinator). Clearly, as discussed by Turner and Turner (1982), liminality here is a socially unifying experience. The common cause for celebration, and the common experience of a transition marked by a shared ritual, together create a sense of community among the participants.

Liminal space and liminality in space

Van Gennep (1960) argued that liminality is a vitally important part of the rite of passage. In the Gold Coast during Schoolies Week, elements of liminality adhere to the place, and also to the carnivalesque loss of identity (Shields 1991: 97) which is served by anonymity and by experimentation with newly intensified bodily experiences. Though, for many, the celebration through clubbing, consuming alcohol, drugs and sex is not a new experience, in this place it can be practised unleashed, and it is compressed in time and space into a few days in the heart of Surfers Paradise. The GCCC spokesperson noted that the coastal setting, the hotels and infrastructure, and the life-style make the Gold Coast the 'logical' place for such a ritual to take place. This was echoed in some interviews: 'there was no way we were going to stay home for Schoolies, so, the Gold Coast' (Anne and friend); 'it's just permanent fun up here' (Rebecca).

Cavill Mall, known simply as 'Cavill' to many Schoolies, turns into Schoolies' territory during the Schoolies Week period. It exemplifies the role of place liminality in allowing identity to become fluid in the course of the transition from child/adolescent to young adult. The Mall, which leads directly to the main Surfers Paradise beach (Figure 4.1), is where rites of passage are most publicly on display. It is where carnivalesque street party dimensions are performed and viewed. Here on the Mall the age and monetary restrictions which keep some Schoolies out of night-clubs are lifted. For the duration of Schoolies, 'Cavill' is transformed, in the evenings at least, into the territory of a Schoolies' street party. Here the spatial order of the dominant culture is suspended as every evening thousands of Schoolies amass on the mall in a blend of participation and spectating:

> Everyone was just going up to Cavill Avenue, that's its own night-club.
>
> (Manager, Islander Hotel)

> A seething mass of humanity, and it's basically fifty per cent of the population watching the other fifty per cent.
>
> (Bruce, event co-ordinator)

In this street party, as in carnival, the dominant social order is inverted and its performance codes are re-articulated. In one sense space and time are frozen as the daily order of things (work, trade, even the holidays of 'ordinary people') are almost overwhelmed by the party:

> They're quite happy to stay on the street and enjoy the moving 'human party' which was free, you could interact with it. They're very territorial, they don't tend to move . . . So, they're not going to move, this passing parade and this cavalcade of face and whatever else makes up Schoolies is so enchanting.
>
> (Bruce, event co-ordinator)

The suspension of time, space and social order produces a liminality which creates an ideal physical space and the ideal conditions for rites of passage to unfold:

> But I think they were astounded when they turned up 'cause their peers or their brothers or sisters had said that 'nothing happens' in the sense of the 'nothing happens' but it's in the nothing that happens that is the happening, you know what I mean, so badly put, but it's really what they come for.
>
> (Bruce, event co-ordinator)

After the years of childhood/adolescence ordered by rules and regulations and the structured organisation of school life, the 'nothing happening' is itself ritually celebrated on the Mall as 'the happening' and as an essential part of the rite of passage to maturity.

One indication of the suspension of the normal rules is the temporary relaxation of the regulation of behaviour by controlling agents such as the police. While there was a crack-down on the entry of Schoolies under 18 years into night-clubs and bars, public drinking was treated with ambivalence. One Schoolie recounted an incident in which he was fined $38 by police in Cavill Mall for drinking in public. However, on learning that he was under-age, the fine was replaced with a warning. The GCCC sponsored a number of Chillout Zones around the Mall and on the beach where Schoolies who were distressed, disoriented, drunk or drugged could come to rest, get cleaned-up and be collected by their friends or returned to their accommodation, sometimes by the co-ordinating police officer.

The very high presence of police during Schoolies Week was maintained as unobtrusive. Despite police walking the length of the Mall in groups of four or five every thirty minutes, Schoolies interpreted this as protective rather than restrictive. No Schoolies interviewed made negative comments about the police presence. One Schoolie commented:

REBECCA: They've been pretty protective. Like they're around all the time.
INTERVIEWER: You've noticed that.
REBECCA: Yeah there's a lot of them out there. We counted twelve walking down the street the other night . . . they're not picking on anyone. You feel heaps worse in Sydney . . . They're just keeping their eye on it. They're much better than our police. They haven't restricted anything.

The collusion of city and coercive authorities in the creation of space in which liminal behaviour is accommodated can however be seen as emphasising the dominance of the hegemonic social structure which will be re-established after the event (Seebohm 1994).

Control and rejection

In 1996 the GCCC took on what they saw as a 'dysfunctional event' and developed a five-year strategy of regaining 'credibility' of Schoolies Week and rehabilitating it as a 'Schoolies Festival' under the themes of 'Celebration and Control' (Alan Russell, GCCC). This reclaiming of the 'event' was aimed at recovering the image of Schoolies Week (and its implications for the tourist image of the Gold Coast), and at generating repeat visits from Schoolies and their own families in later years. Council hired an event co-ordinator and organised all-ages alcohol- and drug-free activities, outdoor movies, band performances, beach volleyball, beach football and street markets, all of which were considered 'strategic diversions' for Schoolies. A public relations company was engaged to manage the Festival's image and control access to sources of information regarding Schoolies (e.g. arrest statistics). This company, for instance, would release no information to the field researchers until detailed discussion with the authors on the purpose and intent of the research had been held.

Eco (1985) discussed carnivals and other similar events as 'authorised transgressions' rather than real transgressions because they take place in an authorised and controlled environment. Because of the location of transgression within an authorised framework and in designated spaces, they are reminders of the existence of rule. The strategic diversions organised by the GCCC can be interpreted in this way. The events, for instance, aimed to rein in the excesses of Schoolies' behaviour by channelling their movements and directing their transgressions:

> Our initial involvement was to take what was a haphazard, randomly happening event and give it some sort of structure, but not structure so that it turned into a summer boot-camp for kids . . . The programme and events has a twofold purpose, one is to add to the celebration but secondly they're there in terms of a strategy for controlling or monitoring crowd movements through an area which is pretty tight.
>
> (Bruce, event co-ordinator)

Beach activity, in a space normally not incorporated into the system of controlled civilised spaces due to its association with the natural world, recreation and liminality, is increasingly being incorporated into a more structured approach to Schoolies Week.

Jackson (1988: 220) discusses the threatening nature of carnivalesque transgression as deriving in part from particular forms of unrestrained 'masculine' behaviour. The GCCC's spokesperson discussed the attempt to 'feminise' Schoolies Week in the attempt to alter its public image and contain what was construed as threatening 'masculine' forms of behaviour:

> We would like to create activities like markets and all sorts of weird and wonderful things that might be classified outside the Australian testosterone-driven male activities which was 'drink as much beer as possible, do as many naughties as you possibly could and beat as many coppers up' and you know if you've had a lot of scratches on the blackboard, you've had a pretty good Schoolies. Well we've certainly tried to move away from that.
>
> (Alan Russell, GCCC)

Council's temporary rearrangement of dominant social relations through containing and channelling Schoolies' potentially transgressive behaviour is an attempt to sanitise these celebratory behaviours and locate them as temporary, understandable and acceptable departures from normative social ordering. The

Council's attempts at control are mirrored in increasingly strict regulation of behaviour by hotel owners aiming to ensure that Schoolies' excesses are restricted to the public spaces of the Mall and the beach. For example, some hotel owners segregated Schoolies, in some cases from each other and in others from the other guests, one hotel would only put Schoolies in fully tiled units to ease cleaning, and another only allowed female Schoolies to stay. Many hired extra security during Schoolies Week.

The liminality of Schoolies Week, that essential component of a rite of passage which creates a space of discontinuity in the social fabric, may be destroyed by the 'celebration and control' campaign of Gold Coast authorities. The temporary authorisation of limited and contained transgression, bringing into the mainstream previously marginalised activities, may alter the function of Schoolies Week as a rite of passage, a symbol of transition to a new phase of life. In this sense, the authorisation of the transgression deletes the transgression.

There is some evidence that Schoolies Week at the Gold Coast is losing its popularity as Schoolies seek other, less controlled locations that allow their rituals freer reign. Many interviewees noted other locations where friends had gone to celebrate – Port Macquarie, Coffs Harbour, Noosa, Byron Bay (all on the Australian east coast), as well as Fiji and Bali. Schoolies' numbers in 1996 were fewer than in previous years and there was evidence of a rejection of some of the staged structured activities:

> We ran an under-age night club . . . We organised a venue and so on. It was unsuccessful. It was supposed to run for seven nights, free soft drink, paid entry. We had DJs from every night-club across the Gold Coast. We had security on, advertisements. We got 30 people in, it got canned after three nights. So we tried to do the right thing by them and it didn't work.
>
> (Manager, Islander Hotel)

This may be indicative of the desire to reject the institutionalisation and control of a rite of passage, the essence of which is a celebration of rejecting institutionalised authority.

Conclusion

This rite of passage from youth to maturity exhibits a specific geography which is experienced at different scales; particularly the scale of the individual body and the regional scale of the chosen location. Both the embodied experience and the urban environment incorporate contradictory elements in the transitional process. Schoolies Week is an intensely physical experience in which young bodies celebrate freedom from youthful restrictions imposed by school, parents and familiar locations. However, those bodies are also careful to temper individual bodily excesses with a group concern for health and personal safety. At the individual level, the celebration is tempered by control.

At the urban scale, the Gold Coast is the chosen location for thousands of Schoolies because of its reputation as a holiday location, where theme parks, beaches and sunshine combine to make it a place not only of fantasy and escape, but transition and becoming. Paradoxically this transformative space is focused on Cavill Avenue, a pulsating shopping mall at the heart of Surfers Paradise, where the happening is 'nothing happening'. Conversely, the classically liminal spaces of the beach and foreshore are incorporated into beach parties and chill-out zones, places which are on the edges of transformative space.

Increasingly the celebration of Schoolies Week has been controlled by the GCCC and other agents of hegemonic authority. The carnivalesque rituals performed by Schoolies have in this process of control and appropriation become themselves transformed. The increasing regulation by authorities has shifted the carnival from its former meaning as a transgressive ritual of resistance (Lewis and Pile 1996) to a blander series of staged events which can be seen more as a safety valve for the control of youthful exuberance – a form of 'bread and circuses' (Craton 1982). In so doing, the Gold Coast authorities argue that the carnival has become feminised, from a hyper-masculine rampage of threatening drunken behaviour, to a controlled, quieter and more diverse form of celebration. Moreover, the City Council's increased control allows authorised transgressions suspending the normal social order, although only as a temporary social inversion in a defined space, diffusing resistance and assuring the incorporation of the Schoolies into their new position in the social order once the rite of passage is complete. However, the very essence of Schoolies Week is bodily excess and lack of structure. The attempt to increase control may well threaten its role as a rite of passage. Already Schoolies are seeking other less controlled locations in which their celebratory behaviours will, at least for a while, retain their carnivalesque marginality in settings where authorities do not yet attempt to interfere, sanitise and control.

References

Bakhtin, M. (1968) *Rabelais and his World*, Cambridge, Massachusetts: MIT Press.

Craton, M. (1982) *Testing the Chains: Resistance to Slavery in the British West Indies*, Ithaca, NY: Cornell University Press.

Dorn, M. and Laws, G. (1994) 'Social theory, body politics and medical geography', in *The Professional Geographer* 46, 1: 106–110.

Eco, U. (1985) *Truths and Transgression*, New York: Harper.

Fiske, J. (1989) *Reading the Popular*, Boston: Unwin Hyman, 43–76.

Herod, A. (1993) 'Gender issues in the use of interviewing as a research method', *The Professional Geographer* 45, 3: 305–317.

Jackson, P. (1988) 'Street life: the politics of Carnival', *Environment and Planning D: Society and Space* 6, 2: 213–227.

Jones, M. (1986) *A Sunny Place for Shady People: The Real Gold Coast Story*, Sydney: George Allen and Unwin.

Jorgensen, D. (1989) *Participant Observation: As Methodology for the Human Sciences,* Newbury Park, California: Sage.

Lewis, C. and Pile, S. (1996) 'Women, body, space: Rio Carnival and the politics of performance', *Gender, Place and Culture* 3, 1: 23–41.

Longhurst, R. (1995) 'The body and geography', *Gender, Place and Culture* 2, 1: 97–105.

McKendrick, J.H. (1996) 'Back to basics: epistemology and research design in multi-method research', in J.H. McKendrick (compiler) *Multi-method Research in Population Geography: A Primer to Debate*, Population Geography Research Group (of RGS with IBG) and The University of Manchester, 46–56.

Minichiello, V., Aroni, R., Timewell, E. and Alexander, L. (1995) *In-depth Interviewing: Principles, Techniques and Analysis*, Sydney: Longman.

Mullins, P. (1993) 'Tourism urbanisation', *International Journal of Urban and Regional Research* 15, 3: 326–342.

Myerhoff, B. (1982) 'Rites of passage: process and paradox', in V. Turner (ed.) *Celebration: Studies in Festivity and Ritual*, Washington DC: Smithsonian Institutional Press, 109–135.

Rojek, C. (1985) *Capitalism and Leisure Theory*, London: Tavistock.

Schoenberger, E. (1991) 'The corporate interview as a research method in economic geography', *The Professional Geographer* 43, 2: 180–189.

Schouten, J.W. (1991) 'Personal rites of passage and the reconstruction of self', *Advances in Consumer Research* 18, 2: 49–51.

Seebohm, K. (1994) 'The nature and meaning of the Sydney Mardi Gras in a landscape of inscribed social relations', in R. Aldrich (ed.) *Gay Perspectives II: More Essays in Australian Gay Culture*, Sydney: University of Sydney, 193–122.

Shields, R. (1991) 'Ritual pleasures of a seaside resort: liminality, carnivalesque, and dirty weekends', in *Places on the Margin: Alternative Geographies of Modernity*, London: Routledge, 73–116.

Spradley, J. (1980) *Participant Observation*, New York: Holt Reinhart & Wilson.

Turner, V. and Turner, E. (1982) 'Religious celebrations', in V. Turner (ed.) *Celebration: Studies in Festivity and Ritual,* Washington DC.: Smithsonian Institutional Press, 201–216.

Van Gennep, A. (1960) *The Rites of Passage*, trans. M.B. Vizedom and G.L. Caffee, London: Routledge and Kegan Paul. First published in 1909, *Les Rites de Passage*, Paris: Noury.

Winchester, H.P.M. (1996) 'Ethical issues in interviewing as a research method in human geography', *Australian Geographer* 2, 1: 117–131.

——(1999) 'Interviews and questionnaires as mixed methods in population geography: the case of the lone fathers in Newcastle, Australia', *The Professional Geographer* 51, 1: 60–67.

5

PREGNANT BODIES, PUBLIC SCRUTINY

'Giving' advice to pregnant women

Robyn Longhurst (New Zealand)

Pregnancy is an important 'rite of passage' (Kimball 1960). One facet of this rite of passage is pregnant women giving, and, more usually, receiving advice. In particular, women who are pregnant for the *first* time seem to receive an enormous amount of advice as they make the transition to motherhood. This advice comes from a range of people including health workers, friends, acquaintances, loved ones, colleagues and even strangers. It also comes from a range of sources including pamphlets, manuals, books, newspapers, magazines and advertisements. The advice often focuses on topics such as diet, exercise, birth, medical procedures, lactation, how to care for a new-born baby and how to raise a child. Although this battery of advice is to an extent welcome and regarded positively, many pregnant women also experience a sense of being under surveillance, and of being regarded as vessels for a foetus whose well-being is the primary object of the advice-givers. The 'giving' of advice can be read as helpful but it can also be read as an attempt to impose limits on pregnant women's bodies and behaviours. I focus on this latter reading.

I outline below the methodological process that was engaged to carry out the research. Data were collected from thirty-one pregnant women who were pregnant for the first time and were living in Hamilton,[1] Aotearoa/New Zealand.[2] I move on to explain that pregnant women are often constructed as being 'in a condition' and, therefore, in need of a great deal of advice. Examples of the advice they often receive are provided. I also examine a 'Birth Exposition' that was held in Hamilton in 1993. Third comes a discussion of the particular role played by pregnant women's husbands/male partners in relation to giving advice. Fourth, I focus on the touching of pregnant women's stomachs. Fifth, I argue that the New Zealand Plunket Society has played a vital role in keeping pregnant women under surveillance through the giving of advice. Finally, I argue that sometimes when pregnant women refuse to take notice of advice they are constructed as antagonistic towards their foetus.

The research which I discuss here was conducted over a period of approximately two years – May 1992 through to July 1994. In the discussions with the participants, we talked generally about their experiences of pregnancy, as well as the advice they received, and sometimes passed on to others. I asked the women questions such as: 'What activities have you continued to carry out during pregnancy and what activities have you reduced or stopped carrying out during pregnancy?' and 'Which places have you continued to visit during pregnancy and which places have you reduced or stopped visiting during pregnancy?' In the course of asking pregnant women questions such as these they inevitably discussed advice that had been given.

I conducted focus groups (see Longhurst 1996a), interviews and in-depth (ethnographic) work for this research. Part of the in-depth work with four pregnant women involved them keeping a journal during their pregnancy. They each gave this journal to me after they had given birth. I asked all the pregnant women about their experiences of public places, for example, work places, leisure places, shopping places. I also spoke with some of these women's husbands/male partners, and with Hamilton midwives. During the final stages of the project I myself became pregnant (with my second child). I also draw on this experience in the chapter.

In a 'condition' – in need of advice?

One of the major findings of the study was that most of the pregnant women with whom I spoke tended to withdraw from public places during pregnancy (Longhurst 1996b). One possible reason for pregnant women going out less is that a number of these women constructed themselves, and felt as though they were often constructed by others, as naturally anarchic and disordered in their thinking and behaviour during pregnancy (Longhurst 1997). There is a discourse that pregnant women tend to be more emotional, irrational and forgetful than non-pregnant women, and than men. This bodily and mental 'difference' functions to disqualify them from stepping 'objectively' and 'dispassionately' into the public sphere long associated with 'Rational Man'. Perhaps it ought to come as no surprise then that these pregnant women – their disorderly bodies and minds – were widely considered to be in need of a great deal of advice.

Pregnant bodies are also often considered to be in a physical 'condition'. Some health professionals continue to define pregnancy and other reproductive functions as requiring medical treatment (see Young 1990: 168–169, citing Katz Rothman 1979: 27–40). This is despite the fact that women often have 'a sense of bodily well-being' and an 'increased immunity to common diseases such as colds, flu, etc' during pregnancy (Young 1990: 170). This tendency to treat pregnancy as a 'condition' can lead implicitly to a conceptualisation of women's reproductive processes as disease or infirmity. Treating pregnancy as a 'condition' can also lead to pregnant women receiving lots of advice.

There are many examples of advice often given to pregnant women that I could draw on in this chapter. However, I have chosen to offer just four – advice about diet, advice from male colleagues about ultrasound scans, advice about taking it easy and advice on a range of topics given at a three-day Birth Exposition. I discuss each of these examples in turn.

It is very common, on having pregnancy confirmed by a health professional or during a prenatal first visit to a physician or midwife, for the 'becoming mother' to receive advice on diet – what the body takes into itself. This advice may be verbal and/or written. In my first visit to the midwife I received verbal advice as well as a number of pamphlets such as 'Listeria in Pregnancy',[3] 'Food Fantastic' and 'Iron in Pregnancy: Nutrition for Two'.

Sheila Kitzinger, one of the foremost pregnancy and childbirth educators (advisers?) of the past two decades, discusses the nutritional needs of pregnant women in depth. In her well-known book *Pregnancy and Childbirth* (1989: 89) Kitzinger recommends that pregnant women have 92 grams of protein a day for optimum health (this is twice as much as for women who are not pregnant). Also, milk is recommended for the pregnant woman. Instructions are also given concerning the intake of carbohydrates, fats, vitamins and minerals. Interspersed with these instructions are many warnings about the dangers of putting on unnecessary weight – 'Cakes, puddings and biscuits do not do much to help your unborn baby's health. If you like sugar in tea and coffee, train yourself to enjoy both of these without it' (Kitzinger 1989: 89). (See also Swinney 1993, cited in Bell and Valentine 1997: 46.) Pregnant women are also advised not to drink alcohol, smoke tobacco or take unprescribed drugs or medication.

Yet, and perhaps it is not surprising, some pregnant women contest and/or resist such advice. Rebecca provides a useful example. Rebecca's colleagues advised her against eating potato chips and spicy foods. She retorted 'I'll eat potato chips if I want to . . . I'll just eat normally. So!' Rebecca also continued to smoke despite being advised many times, and by many people, to quit. She explains 'It's all very well to say "stop", but you can't!'

A second example of advice that is commonly given to pregnant women is that they should have at least one ultrasound scan in order to predict as accurately as possible the date on which the baby might be born and to check that 'things look normal'. Kerry worked training people in telecommunication systems and was the only woman working with a group of about twenty men. I visited and talked with Kerry at least once a month throughout her pregnancy. She explained to me that when she became pregnant her male colleagues frequently gave her advice, including advice on ultrasound scans:

> They were all coming out with these things about what you've got to do and how much extra for a video and they know all the stuff. It was really funny. They were going 'make sure you drink two litres' and then another guy said 'no, you don't need to drink two litres, you only have to drink half a litre'.

The comments from these men could be read in a multitude of ways. The comments could indicate how much more men have to do with birth now than in previous years. I suspect these conversations would not (and could not) have occurred in work places twenty years ago when pregnant women were not as visible in the work place, and when men did not routinely accompany their partners on visits to health workers, antenatal classes or the birth itself. Another reading of these men's comments could consider the contestatory nature of their claims to 'true' knowledge about the processes of pregnancy, birth and lactation.

KERRY: But it's actually quite funny all the different advice they give you sort of thing. They're telling me they can tell what size baby you're going to have 'cause of the measurement of the skull. And one was arguing 'no, it's not the measurement of the skull, it's the measurement of the chest'.

Their comments could also be read as laymen co-opting the processes of birth similar to the way in which professional men took control and medicalised birth in industrialised nations several decades ago. Others may read their comments as sexual harassment. Yet, for the purposes of this particular argument I read the comments of these men as *advice* given to a pregnant colleague. They felt it important to inform Kerry about something they themselves felt knowledgeable about. Kerry's response to her male colleagues offering her advice (one colleague even advised her to put her husband's handkerchiefs down her bra when she started breast feeding so that she wouldn't get stains on her clothes!) was to treat it light-heartedly. 'It was really hilarious', Kerry said. I expect, however, that at times the hilarious may have bordered on the tiresome and the irritating as Kerry also made comments such as 'I mean, really!'

A third example of advice that pregnant women often receive from colleagues, as well as from friends, family members and loved ones, is to rest and sit or lie down as much as possible, to not 'overdo things', to not lift anything heavy or stretch and bend too vigorously and to not partake in some sports such as running, skiing, diving and horseback riding (see Longhurst 1995). Denise's doctor also told her not to go bungy jumping. Denise's friend Kerry, who was also pregnant and present during this conversation responded by saying 'You wouldn't do it [bungy jumping] anyway, would ya?' Denise replied: 'Yeah, I would, I would. I want to do that. That's one thing I want to do. I don't think it would be that bad'. In fact, Denise did not go bungy jumping during her pregnancy but she did not take kindly to the doctor telling her that she should not do it. Being told that it was 'off limits' seemed to make Denise want to do it more rather than less.

Most women in the study were not in the least bit motivated to go bungy jumping. They did want to get on with tasks such as lifting heavy objects, gardening, home decorating and continuing to exercise, but their desires were often met with opposition:

DONNA: I wasn't allowed to do any lifting. I got my head snapped off if I even just thought about it.

SAM: I know that from the moment that my grandparents found out that I was pregnant it was almost like, you don't do any activity at all; you sit round with your feet up all the time, you rest and you have afternoon sleeps.

JILL: Mainly, you know, people say to me 'sit down and put your feet up'. I say 'I can't. I want to get this wall-paper stripped'.

So far I have discussed several topics on which first time pregnant women are often advised (diet, ultrasound scans and levels of exertion). Now, however, I want to discuss a particular place where pregnant women receive advice.

It is common for pregnant women to receive advice at places such as doctors' surgeries, midwives' clinics, home birth clinics, hospitals and antenatal classes but in 1991 and 1993 many pregnant women also received advice at the Hamilton Birth Exposition (which is commonly referred to as the 'Birth Expo') held at the Hamilton Gardens Pavilion. I attended the Birth Expo in 1993 which was held over a period of three days, 31 April to 2 May. Those attending were required to pay a $2.00 entrance fee at the door.

There were approximately thirty organisations represented at the Birth Expo.[4] Advice was offered by way of pamphlets, displays, photos, videos, models, large posters and people to talk with. There were also products, such as Weleda (naturopathic and homoeopathic medicines, oils, remedies) and Tetra (baby bedding) available for purchase.

All this advice was disseminated by women, many of whom themselves had babies close by. Most of the 'consumers' of this advice were also women, some of whom had male partners and children accompanying them. In terms of 'race'/ethnicity and social class, both those who were representing organisations or selling products and those attending the Birth Expo appeared to be Pakeha[5] and reasonably 'well off'. There were perhaps 100 people in attendance at the particular time that I was there (2–3 p.m. on Sunday, the final day of the Birth Expo), and at least ten of those participants were visibly pregnant.

My reading of the cultural landscape of this Birth Expo is multiple and ambivalent. On the one hand, birth is 'exposed' – brought out into the public arena for discussion. The pregnant women who attend are subjects who exercise their agency by choosing to pay $2.00 in order to gather and possibly swap advice. In this regard, the Birth Expo fulfilled an important and positive function. On the other hand, the Birth Expo provided the perfect arena not just for helpful advice but for the policing of pregnant women's behaviours. Pamphlets aimed at selling products, for example, 'Safe T Wraps' for babies frequently contain slogans such as 'Peace of Mind for Caring Parents', the implication being that if you do not buy the particular product, you are not a caring parent. The words 'Cot Death' directly underneath the aforementioned slogan serve to reiterate the message that if you do not buy this product you are not a caring parent, and you may even risk your baby's life.

There are also 'information' sheets on how you must take precautions during pregnancy in order to 'prevent irritation of joints', for example. There are 'essential pelvic floor exercises' and the stretch classes that you 'should' attend during pregnancy in order to give you 'a sense of well being and aid relaxation'. The message from the community drug/alcohol resource centre is that 'when you drink, so does your baby . . . stop drinking if you are pregnant or planning a pregnancy'. In this instance the advice given is not only to pregnant women but also to non-pregnant women who might be contemplating conception! Along with this there are advertisements for the usual prenatal courses as well as for 'early pregnancy classes'.

To summarise thus far I have argued that women are often encouraged to see themselves as more disordered in their thinking and as being in a 'condition' when pregnant. They are, therefore, considered to be in need of advice as to how they might best take care of themselves, and even more importantly, of their foetus. This advice does not just come from health professionals, antenatal classes or from older women (as is sometimes thought), rather, it comes from a range of people – even male colleagues. Pregnant women also often receive advice from husbands/male partners. Their advice forms the basis for the next section since the position occupied by husbands/male partners in relation to pregnancy is particularly interesting.

Husbands/male partners: support or growling?

Often the foetus represents not the individual concern of the 'becoming mother' but rather the joint concern of the 'becoming mother and father'. There is a linguistic term currently in vogue in Aotearoa/New Zealand – 'we're pregnant'. It is now part of the dominant discourse that men ought to 'share' in pregnancy. A booklet given out to most pregnant women during their first antenatal visit states: 'Fathers share much of the excitement and worries of pregnancy' (*Your Pregnancy* 1991: 14). It seems that part of this sharing in the pregnancy means offering 'support' to their wives/partners. In another booklet (Parker 1994), which is also distributed by health practitioners to most pregnant women in Aotearoa/New Zealand, it is stated: 'While all this [pregnancy] is understandably bewildering for the man, it is important that he understand what is going on and support his partner'. Yet what support entails is not specified.

The word 'support' is defined in the *Collins English Dictionary* (1979: 1460) as '1. to carry the weight of. 2. to bear or withstand (pressure, weight etc.). 3. to provide the necessities of life for (a family, person etc.)'. Support, according to this definition, does not necessarily entail understanding and respect. In relation to 'becoming fathers', supporting their pregnant partners often seems to mean advising them as to what they should not do. Support becomes conflated with admonishing and cautioning. For example, I talked with pregnant women whose husbands/male partners told them that they should not worry about putting on weight, they should not lift heavy things and they should not paint ceilings. Dorothy claimed: 'I've actually put on quite a bit of weight, but he's never really mentioned that, in fact he's always growled at me when I worry about it' (individual interview). Mary Anne, too, said that she was 'growled at' by her husband:

> Barry did all the heavy lifting, I just did the unpacking and moving of the light stuff and I moved a couple of heavy things when my husband was working and got growled at when he came home and realised that I'd moved them.

Helen said 'Gary went ape [became angry] when he found out I'd painted the ceiling. I suppose perhaps I did overdo it a bit'. Yet in all these instances, the women also described their husbands as caring, supportive and offering advice. Offering advice (read: support) and growling were not considered to be mutually exclusive, in fact, they seemed to be mutually constitutive. Growling was seen as a gesture of caring and support. These women were quite open to having their husbands discipline their behaviours, in much the same way that an adult might growl at a child. As I argued in the first section, women are constructed as needing extra advice, guidance, protection and disciplining during pregnancy. This disciplining happens not only through husbands/male partners growling, it also happens by way of people touching pregnant women's stomachs.

Touching pregnant women

Many of the pregnant women with whom I spoke explained that not only did they receive lots of advice but people also touched their stomachs – not just loved ones or very close friends but also people who would not usually consider touching their bodies. On several occasions when I was working as a university lecturer and my pregnancy was clearly evident, students (whom I did not know very well) touched my stomach. It is not uncommon for some people to take the liberty of placing their hands upon pregnant women's stomachs. While some women claimed that they found this frustrating and disempowering, others appeared not to mind, and some even enjoyed it. When I asked Christine if she had experienced people touching her stomach she replied:

> Yeah, I think people do think they have sort of a right to comment, but most people have been pretty good. It was 'hard-case', like a friend the other day at church, he sort of gave me a little pat, I didn't mind, but it would just depend who it was type of thing.

Sonya, who was 26 weeks pregnant at the time of the interview, was less receptive to people touching her stomach although she herself admits to having touched other pregnant women's stomachs in the past without invitation:

> I used to go up and put my hand on as if it was my property because it stuck out [laughter] but it's not you know, and I know now how personal it actually is. It's nice to have somebody come up and put their hand on as long as it's invited on. I wouldn't like somebody just to come up and shove their hand on or something.

In a journal entry Paula explains:

> Sometimes I feel as though being pregnant automatically deprives me of any individual identity and personal space. People seem to have a fascination with pregnant women's stomachs and want to pat them. It's not something they would normally do, but because I've got a 'bump' it seems that I've become public property. Complete strangers seem to want to be 'involved' in the pregnancy process. I often get stopped in shops (particularly the supermarket) to be asked when I'm due, how I'm coping with the summer heat etc. Then the advice and personal stories start.
>
> (cited in Longhurst, forthcoming).

Young (1990: 160) claims that: 'Pregnancy does not belong to the woman herself. It is a state of the developing foetus, for which the woman is a container'.[6] This has not necessarily always been the case. Pregnant bodies need to be 'located' not just in relation to place but also in relation to time or history. Approximately one hundred years ago in Aotearoa/New Zealand the surveillance of mothers happened not so much during pregnancy itself, but rather, after the baby was born.

New Zealand Plunket Society

In the early 1900s mothers began to receive lots of advice about their newborn babies. In August 1913, Truby King gave an address at the National Congress in London entitled 'The New Zealand Scheme for Promoting the Health of Women and Children'. King (1913: 3) stated that the first aim of the 'Society for the Health of Women and Children' was 'To uphold the sacredness of the body and the duty of health; to inculcate a lofty view of the responsibilities of maternity'. In 1915, King George V conferred on the Society the title of 'The Royal New Zealand Society for the Health of Women and Children'. From around 1925 the Royal New Zealand Society for the Health of Women and Children (Inc.) became more commonly known as the Plunket Society. The Plunket Society aimed at producing healthy children and making infant mortality in Aotearoa/New Zealand one of the lowest in the world. Over the years it has been widely stated that millions of New Zealanders 'owe a debt of gratitude to Sir Truby King' (Snowden and Deem 1951: 8). In *Modern Mothercraft*, which was the official handbook of the Plunket Society, it is claimed: 'The work of the Plunket Society has had a profound effect in laying the foundations for a healthy nation, and there is growing evidence, too, that other countries are finding the system of great help and value' (Deem and Fitzgibbon 1953: foreword).

I want to argue, though, that King's régime did much to put women under surveillance and constructed a cultural hegemony around mothering that had not existed in the same way prior to his interventions (see Sullivan 1995: especially 13–14). Not only was breast feeding (or natural feeding as it was often called) heralded as ideal for all mothers and babies but also strict daily routines were laid out (see Deem and Fitzgibbon 1953: 52–71). The suggested 'Routine Day for the Nursing Mother and her Baby' detailed in the Official Handbook of the Plunket Society (Deem and Fitzgibbon 1953: 63) is testimony to this regimentation – the social inscriptions – that mothers (and babies) in Aotearoa/New Zealand faced during this era.

A discourse involving the norms of corporeality, measurement, technique and judgement quickly grew up around King's notion of the ideal infant (see Sullivan 1995: 14–16). A comprehensive surveillance infrastructure emerged which subjected the mothers and infants to continual scrutiny. One of the forms that this scrutiny and policing took was the issuing of a *Plunket Book* to all mothers from 1924 onwards (Sullivan 1995: 14). These books are still issued to all mothers/infants today in Aotearoa/New Zealand. Plunket Books contain a weight chart, room for mother to record baby's progress, or lack of progress, and space for the Plunket nurse to record her observations and suggestions about feeding and care. It can be nerve-racking for mothers waiting for the Plunket nurse to visit; waiting for her to measure and weigh the baby, wondering what she will enter into the book – an entry that will be recorded for posterity.[7] Moana, a Maori woman in her early twenties explained to me that she wanted nothing to do with the Plunket Society after the birth of her baby. She was visited by a Pakeha Plunket nurse once but did not feel comfortable. Moana felt as though the nurse did not approve of or respect the type of care that she was offering her new son, for example, sleeping in the same bed with him.

Today, however, the giving of advice, surveillance and policing, happens much earlier: it does not just happen once the baby is born but rather as soon as a woman is 'confirmed' pregnant. Currently, not only is a record kept of the baby's progress after birth, but also a record is kept of pregnant women's health, 'well being' and 'growth'. It is usual during antenatal visits for pregnant women to have their urine analysed for the presence of sugar and albumen (protein). They are then weighed and have their blood pressure measured. Then, the fundus (top of the uterus) is measured, the foetal heart is checked and the abdomen is palpitated in an attempt to determine the baby's position. This information is noted by the General Practitioner or Midwife in their records. It is also entered on a card which pregnant women themselves keep. Women bring this card with them each time they have an antenatal check-up. This record functions in a very similar way to the Plunket Book. The record can be seen as a surveillance tool which works to regulate and normalise the bodies of pregnant women. Whereas mothers used to be supervised in relation to their treatment of babies and children (usually through the Plunket Society) they are now also supervised in relation to their treatment of the foetus. This raises an important question: what if pregnant women refuse to heed advice about how to best look after their foetus?

Refusing to heed advice

In the USA there have been a number of attempts made in recent years to penalise women for activities undertaken during pregnancy – activities believed to be harmful to the foetus, activities that women have been advised to avoid. Chavkin (1992: 195–196) cites some examples.

In 1980, a Michigan court held that a boy could sue his mother for taking antibiotics during her pregnancy, allegedly resulting in the discoloration of the child's teeth. Another Michigan court decided that evidence concerning a woman's 'prenatal abuse' of her foetus could be obtained by reviewing her medical records without her consent. A well-known such case is that of Pamela Stewart, a San Diego woman who faced criminal charges for not following doctor's orders to stay off her feet during pregnancy, abstain from taking amphetamines and summon medical assistance when she went into labour.

What these cases illustrate is that not only are pregnant women/foetuses increasingly under surveillance in order that specific codes of behaviour are adhered to (that advice is followed), but also, that pregnant women are often thought to be in battle against their foetuses if they do not follow advice. If a pregnant woman fails to act in accordance with advice and if 'she deviates from medically, socially, or legally sanctioned behaviour' (Chavkin 1992: 193) then she is likely to be positioned as antagonistic to the foetus. While such legal precedents have not been set in Aotearoa/New Zealand, I think that the discourse of pregnant women as somehow errant and in need of advice is clearly evident.

Conclusion

In this chapter I have not discussed how some first time pregnant women seek out and make discerning decisions about advice they are given. Nor have I discussed the passing on of advice from one pregnant woman to another, or from older women to younger women. Currently there are competing discourses around pregnant women drinking alcohol, taking non-prescription drugs and/or smoking tobacco. I have not been able to fully engage with these debates in the chapter. Nor have I discussed advice about 'natural' pregnancy and births which is often underpinned by a feminist politic. Nor have I been able to tease out some of the possible differences amongst pregnant women in relation to social class, ethnicity, sexuality and the places they live. There is only so much one can say in a few thousand words! What is evident, however, is that a great deal of research remains to be done in these areas.

What I have been able to argue in this chapter is that pregnant women are represented popularly as being in a 'condition' and not suited to the rigours of sport, physical work, etc. Health practitioners, friends, colleagues, employers, loved ones, family members, husbands/male partners and even strangers frequently offer 'support' and advice which can serve to disempower and reduce pregnant women's autonomy. The behaviour of pregnant women is frequently policed. People often regard themselves as societal supervisors of pregnant women's behaviour. Some even touch pregnant women's stomachs as a way of looking after that property, that potential citizen, in which there is a collective interest. The individual pregnant woman's capacity is primarily as a vessel, while the foetus has a positive and public identity. Yet this supervision was not treated entirely negatively by the participants. Many understood themselves to be making an important journey, a rite of passage, into motherhood, and that receiving lots of advice was 'part of the deal'.

To end on a more general note, however, I want to reiterate that pregnancy is a biological process – but a biological process that always exists within social, cultural, economic and political realms. Pregnant bodies are also always temporally and spatially located (Longhurst, forthcoming). I have used the experiences of pregnant women who live in Hamilton, Aotearoa/New Zealand in the mid-1990s as a basis for this discussion but these experiences are by no means universal. Pregnant bodies are culturally, sexually, ethnically, spatially and temporally specific. They are mutable in terms of their cultural production. In other places, other rites of passage exist for pregnant women. Pregnant bodies are not fixed or stable. The rites of passage that construct and recreate them to perform particular kinds of tasks change over time and place. My hope is that the geography presented in this chapter (along with the other chapters in this book) will help spur on more geographies that also address issues to do with places, bodies and rites of passage.

Acknowledgements

I would like to thank the research participants who so generously gave of their time and stories for this research. I am also appreciative of the conversations about pregnancy that I have had over the years with Catherine Kingfisher, Robin Peace and a number of students at the University of Waikato.

Notes

1 Hamilton is a city of 102,000 people (Census of Population and Dwellings 1991). It is located to the west in the northern half of the North Island of Aotearoa/New Zealand.
2 Aotearoa is the Maori term for New Zealand. Since 1987, when Maori became an official language, the term 'Aotearoa/New Zealand' has become widely used.
3 In the pamphlet published by the New Zealand Department of Health, listeria is described as 'a common bacterium which is found in the soil, water, plants and in the droppings and faeces of animals and humans'. Sometimes listeria can cause a rare illness related to eating contaminated food. This infection is called listeriosis. Listeriosis is considered to be dangerous for pregnant women, as it can cause miscarriage and stillbirth.
4 The organisations represented included the New Zealand Family Planning Association, the Waikato Home Birth Association, the Young Women's Christian Association (YWCA), the Society for the Protection of the Unborn Child (SPUC), the Physiotherapy Department of Waikato Women's Hospital, Hamilton Midwives' Centre, Waikato Women's Health Collective, Hamilton Parents' Centre, Hamilton Midwives' Information Service, Department of Health Dietary Advice, Natural Family Planning and the Hamilton Community Drug/Alcohol Resource Centre.
5 Pakeha is the Maori term for 'A person of predominantly European descent' (Williams 1971: 252).
6 Young's claim has been derived from the work of Kristeva (1980: 237) who argues that 'the mother is simply the site of her proceedings'.

7 When my first son was born in 1991 and I received a Plunket Book for him, my mother presented me with my own Plunket Book which she had kept in a drawer for 29 years. At the same time my partner's mother gave his Plunket Book to him.

References

Bell, D. and Valentine, G. (eds) (1997) *Consuming Geographies: We Are Where We Eat*, London: Routledge.

Census of Population and Dwellings (1991) *Waikato/Bay of Plenty Regional Report*, Wellington: Department of Statistics New Zealand.

Chavkin, W. (1992) 'Woman and fetus: the social construction of conflict', in C. Feinman (ed.) *The Criminalization of a Woman's Body*, New York: Haworth Press, 193–202.

Collins English Dictionary (1979) Glasgow: William Collins Sons and Co.

Deem, H. and Fitzgibbon, N.P. (1953) *Modern Mothercraft: A Guide to Parents*, Official Handbook, Royal New Zealand Society for the Health of Women and Children (Inc.) Plunket Society: Dunedin.

Katz Rothman, B. (1979) 'Women, health and medicine', in J. Freeman (ed.) *Women: A Feminist Perspective*, Palo Alto, California: Mayfield Publishing, 27–40.

Kimball, S.T. (1960) 'Introduction', in A. van Gennep *The Rites of Passage*, London: Routledge and Kegan Paul, v–xix.

King, F.T. (1913) *The New Zealand Scheme for Promoting the Health of Women and Children*, reprint of an address given at the National Congress, London, August.

Kitzinger, S. (1989) *Pregnancy and Childbirth*, London: Doubleday.

Kristeva, J. (1980) *Desire in Language: A Semiotic Approach to Literature and Art*, Oxford: Basil Blackwell, 237–269.

Longhurst, R. (1995) 'Discursive constraints on pregnant women's participation in sport', *New Zealand Geographer* 51, 1: 13–15.

—— (1996a) 'Refocusing groups: pregnant women's geographical experiences of Hamilton, New Zealand/Aotearoa', *Area* 28, 2: 143–149.

—— (1996b) 'Geographies that matter: pregnant bodies in public places', unpublished D.Phil. thesis, University of Waikato.

—— (1997) 'Going nuts': re-presenting pregnant women, *New Zealand Geographer* 53, 2: 34–39.

—— (forthcoming) ' "Corporeographies" of pregnancy: "bikini babes" ', *Environment and Planning D: Society and Space*.

Parker, J. (ed.) (1994) *Baby on the Way*, Hastings, New Zealand: Infant Times Association.

Snowden, R.F. and Deem, H. (1951) *From the Pen of F. Truby King: Chapters Compiled from the Writings and Lectures of the Late Truby King*, Auckland: Whitcombe and Tombs.

Sullivan, M. (1995) 'Regulating the anomalous body in Aotearoa/New Zealand', *New Zealand Journal of Disability Studies*, 1: 9–28.

Swinney, B. (1993) *Eating Expectantly: the Essential Eating Guide and Cookbook for Pregnancy*, Colorado Springs, Colorado: Fall River Press.

Williams, H.W. (1971) *A Dictionary of the Maori Language*, 7th edn, Wellington: A.R. Shearer Government Printer.

Young, I. (1990) *Throwing Like a Girl and Other Essays in Feminist Philosophy and Social Thought*, Indianapolis: Indiana University Press, 160–174.

Your Pregnancy/To Haputanga Me To Whakawhanautanga (1991) Department of Health New Zealand.

6

BODILY SPEAKING

Spaces and experiences of childbirth

Scott Sharpe (Australia)

The practice of childbirth in the West has been characterised by a long history of contestation. Arguments concerning the most appropriate primary care-giver to the birthing woman, and questions relating to authority over decision-making in birth, suggest that childbirth is an intensely political issue. If childbirth has raised issues about the effects of power on bodies, discussion has often been generated by feminists concerned to link childbirth to a history of struggles concerning the medico-technical control of women's bodies.

Geographical issues of childbirth

In his polemic in the early eighteenth century Johan Ettner has doctors plotting to gain access to the birthroom to discover its secrets and to displace the midwives working there (Tatlock, cited by Pringle 1998). As well as highlighting the long history of gendered struggle over the birthing process this story indicates the importance of the use of space to contest the relations of power that exist in regard to childbirth. This territorial element – but not only this, as I will argue later – makes childbirth an important geographical issue. Much contention with regard to childbirth in the West has centred around the problem of *where* childbirth should occur. The interdependence of space, the control of practices and the autonomy women have over their own bodies is not a new issue in geography. Rose (1993) has raised the issue of *confinement*, referring to it as the hallmark of women's experience of space. The very word 'confinement' obviously has specific implications for childbirth in those cases where women have been bound into certain positions, and when their choice of where to birth has been restricted.

Childbirth is, of course, not a geographically uniform event but has variations both between and within cultures and at many geographical scales. International variations in birth have featured in much analysis of childbirth to date. For example, sociologists of health such as Taylor (1979) have used international comparisons of perinatal mortality rates to argue against over-medicalisation of childbirth in countries such as the USA and Australia. At a more local scale, a number of studies has been carried out to compare home and hospital births, often by feminists concerned to give legitimacy to home-birth practices. What studies at both these scales have in common is a certain methodological orientation, insofar as statistical data (such as perinatal mortality rates and measures of birthweight) have been mobilised for the analysis of, and often political intervention into, current birthing practices. The use of statistical measures has been particularly crucial to the arguments of those feminists who are trying to find practical means of limiting the patriarchal control of women's bodies. Oakley (1992), for example, has conducted randomised control trials to show that the involvement of a midwife throughout pregnancy improves an at-risk baby's birthweight.

However, the question of methodology in feminist and other interventions in childbirth debates is itself a contentious issue. Saul (1995), for example, is concerned by the evaluation of childbirth in terms of risk; research such as Oakley's, she argues, legitimates the ordering of a diversity of women's experience in equivalent terms for the purpose of measuring, prediction and control. This chapter distances itself from those methodologies which are based on the massification of differences amongst childbirth practices; that is, it distances itself from the categorical grouping and ordering of birth experience. My methodology is geared, rather, towards capturing the particularity of individual experiences of birth, so that the question of what constitutes a significant and successful birth is not decided in advance but is born out of the singularity of each birth event. On the one hand, the theorisation of the particularity of birth, rather than its generality, requires a very 'local' focus: especially important in this respect are my close attention to the narrations of birth experiences – i.e. the way they are narrated – and my focus on the scale of the body. At the same time, this challenge to assumptions as to what constitutes a significant and successful birth requires that broader questions relating to the conceptualisation of space be considered. I argue in this chapter that the way in which space is conceptualised is closely linked with social practice.

From a geographical perspective, childbirth becomes a particularly interesting and fertile issue when it is recognised that intellectual debates, the spaces in which birth takes place, and the practices that occur there, are by no means independent entities. Indeed, to treat them as clearly independent objects of analyses is to continue to divorce conceptualisation from practice. Furthermore, I challenge the common assumption that lived space necessarily precedes discourse. This challenge, I feel, enables a more productive, and perhaps less oppressive, relation between space, practice and discourse. In showing that space is discursively constructed, I want to demonstrate the interdependence of language and experiential space.

The notion that space is discursively constructed is not a new one in geography. Gibson-Graham (1997), for example, outlines the consequences of failing to recognise space as discursively constructed, arguing that a failure to realise the way language works upon us can reinstate oppressive dichotomies. Employing Sharon Marcus's work on the discourse surrounding rape, Gibson-Graham challenges the notion that 'lived' space is primarily material or *pre*-discursive. The discourse surrounding the 'reality' of rape has led to women being positioned as 'endangered' and 'violable' with the whole female body being symbolised by the 'delicate' inner sanctum of the vagina, an empty space which is in need of protection against trespass (Marcus 1992, cited by Gibson-Graham 1997). The 'reality' of this 'rape script' sees women's sexuality as passive, whereas 'the would-be rapist's feelings of powerfulness' are re-enforced (Marcus, cited by Gibson-Graham 1997: 310). Strategies of rape prevention guided by such discourse are aimed at protecting this inner, immanently violable space but do nothing to challenge the so-called 'truth' of women's victimhood. For Marcus then, as indeed for Gibson-Graham (1997: 311), 'lived space is as much discursively as materially produced' and so how one acts in space is dependent on how it is conceptualised.

This chapter acknowledges the significance of the way that space is conceptualised by scrutinising the relationship between language and space. It also affirms the value of the *body* as a site for geographical analysis. Borrowing from the work of Julia Kristeva, I examine the provisional nature of the distinction between the body and language by drawing attention to the way that the body 'speaks' in language.[1] With the above considerations of the body and language in mind, this chapter explores three main points. The first is the issue of how the birth centre space is discursively constructed. Second, I examine how a focus at the scale of the body both confirms and problematises the predominant constructions of the birth centre space. Finally, as explored in the following section, birth centres have, to date, been somewhat narrowly characterised as an *intermediate space*. So, third, I explore possibilities for reconceptualising the space of the birth centre in a less determinate fashion, in order that opportunities for practice may be opened up.

Labour ward/home birth: a space between

This chapter focuses on one particular birth centre: the Birth Centre at The Royal Hospital for Women (RHW), Sydney. Defining a birth centre is, in no way, a straightforward task, given the contested terrain they occupy. By one reading, birth centres signify a strategic use of space: that is, they represent sites of resistance to patriarchal obstetrics (Whelan 1994). An alternative reading is that they are sites of co-option which have arisen to counter the growing home-birth movement (Whelan 1994). Lastly, in the literature that the birth centre offers to its prospective clients, it is characterised as an intermediate space between home and hospital, offering the best of both worlds (see, e.g., *Information about our Birth Centre* 1993; *Philosophy of the RHW Birth Centre* 1992). Regardless of which of these views is more accurate – and perhaps there exists a combination of these elements – birth centres in Australia have at their origin a growing dissatisfaction with the highly medicalised labour ward practices of the 1960s and 1970s.

In relation to this question of where and how to birth, a fierce debate has raged, constructed around the broad poles of natural versus medical, or home versus hospital, birth. While a birth centre employs a separatism which allows room for relatively autonomous birthing practice, it can also be seen as a compromise to the discourse of 'risk' in birthing practice. The positioning of birth centres in hospital grounds acknowledges the fears and anxieties prospective parents and service providers have about the process of birth and caters for the 'just-in-case' scenario. It is not my intention to argue that birth is inherently dangerous or safe nor that those positioned on either side of the debate are dupes or possessors of false consciousness. Rather, my interest is in the way that the debate itself constructs spaces and bodies which clearly affect people's perceptions of birth and of the birthing practices themselves. This is due to the association of the broad oppositions of medical versus natural birth, and labour ward versus home birth, with a series of related dualisms. Table 6.1 lists some of these dualisms drawn largely from the feminist literature on childbirth (Arms 1975; Balaskas 1983; Martin 1990; Clarke 1996; Pringle 1998) and from the documents that the birth centre issues to its prospective clients (*RHW Birth Centre Protocol: Client's Copy* undated; *Positive Choice* undated; *Philosophy of the RHW Birth Centre* 1992; *Information about our Birth Centre* 1993).

The space of the birth centre and the practices that occur there are both constituted by, and constitutive of, these dualist oppositions. These oppositions cannot be viewed in isolation as many are linked with each other in what has been dubbed by Whelan (1994) the 'cascade effect'. For example, the drive for risk elimination in the obstetric management of childbirth results in women being confined, subject to the gaze of medical technology and, of course, to medical procedures if complications arise. Using the rationale of risk elimination, the yardstick of what constitutes a 'complication' has lengthened in the recent history of Australian obstetrics (Harkness 1986; Pringle 1998). Feminists and women's health activists have criticised this rationale that the risks to the few justify the standardised treatment of the many (Pringle 1998).

More recently however, feminists such as Pringle (1998) and Saul (1994) have been critical of this one-way debate of feminists versus patriarchal obstetrics.

Table 6.1 Dualisms that characterise the politics of childbirth[a]

Feminine	Masculine
Natural (drug-free, non-interventionist)	Medicalised (medicated, interventionist)
Midwife	Obstetrician (obstetric nurse)
Home (personal)	Hospital (institutional)
Private	Public
Normal (healthy)	Abnormal (potential complications)
Technique	Technology
Risk minimisation	Risk elimination
Autonomy	Paternalism
The Touch	The Gaze
Dynamic	Confining
Individual	Standardised

Note:

a The first of these oppositions, in particular, can be seen as a binary opposition in the sense that Derrida and many feminists have used the term to indicate the difficulty in this culture, at this point of time, of thinking the opposing terms independently. While the other terms in this table have a significant history of being opposed to each other, for example in the feminist, childbirth and sociology of health literature, their opposition is not so directly metaphysically entrenched. For this reason, the term 'dualism' has been used to cover the oppositions often associated with childbirth.

Saul (1994) is critical of naïve attempts to use anthropological 'evidence' of the success in human history of traditional birthing methods to authorise contemporary western practices of 'natural' childbirth. Such a strategy homogenises women's experience 'on a (presumed) substrate of universal biology', glossing over cultural differences and contexts (Saul 1994: 3). In particular, Saul worries about the universalising idea that women midwives are naturally more caring. She is also concerned about the theoretical baggage that comes with a term like 'natural birth', insofar as feminists have long been critical of the association of women with nature to justify their exclusion from culture. Pringle (1998) claims that, while thinking in oppositional terms has been useful in creating a space from which health activists could speak, there is now the potential for women to be entrenched as victims, especially as it sets up an irresistible monolith of patriarchal obstetrics. One consequence that may follow from this discourse is that birthing mothers are subjected to a new orthodoxy where non-medicalised birth is set up as the only option; another is that women obstetricians can be potentially marginalised by this feminist discourse (Pringle 1998).

Telling experiences

My research involved interviewing a small number of birth centre users and service providers in order to examine how the birth centre space and the bodies therein were constructed in relation to an already established polarised debate. A recursive model of interviewing was used to allow for the uniqueness of informants' experiences to be given voice (Minichiello *et al.* 1995). This recursive model of interviewing entails a relatively unstructured, open conversation where the interviewer uses 'prompt' words from the informant's discussion to direct it over areas of interest.[2] The purpose of the interviews was not to generalise about people's experiences but rather to attempt to theorise particularity. The transcripts below are not meant to be examined in terms of their representation of a prior, pre-discursive reality, but rather for how they affect and are affected by established debate. The transcripts are also examined in relation to the telling of the experience itself, for reasons which become clear: the narrations, then, have been edited as little as possible. The following transcript shows how the dualisms listed above come through in people's narration of their experiences and indicates the importance of these dualisms for opening up and creating birth centre space:

> I liked it was a separate space from the rest of the hospital even though it wasn't a bad experience being in the labour ward it was still much nicer to be . . . to be able to go through the birth centre . . . to be away from that whole clinical hospital environment where there were strangers walking around all the time and you know a different nurse would come in all the time and just routinely check . . . um my signs . . . you know whatever and . . . um . . . and not really look at me or talk to me much and . . . and it just felt like a patient I suppose whereas with this birth centre it was like being, going into a . . . a little community of women and I felt so confident with this birth that it didn't worry me at all to be separate from the hospital.
>
> (J, RHW Birth Centre client)

In this transcript J can be seen to be championing several of the latter terms listed in Table 6.1; notably, individual, personal, private and feminine. Her reference to her sense of confidence and her not being worried by being 'separate from the hospital' indicates that, for J at least, there is a correlation between an autonomous space and a personal sense of autonomy. Here we see the very real value of the birth centre discursively constructed by these terms as it provides an alternative to the standardised, public, institutional practices of the labour ward. However, the following narration indicates the dangers of championing the latter terms of the oppositions. It shows that by establishing the space of the birth centre as a sanctuary away from the paternalism of obstetrics (manifest in such medical interventions as the administering of anaesthesia such as nitrous oxide or epidural anaesthetic), there is always the danger that a new orthodoxy or conservatism will arise:

> I wasn't actually very happy with that relationship (with the midwife) because she should have examined me sooner and . . . I wanted nitrous oxide and she basically blocked me getting it saying that they'd run out, their cylinder was empty . . . she should have gone and got it for me and . . . um . . . yeah . . . when she said, No there was no nitrous oxide well then because I knew in my mind I couldn't keep going as I was . . . She thought I was just in first stages of labour and I was actually in transition and I'm in thinking, 'I'm going to die if this keeps going' and she's thinking, 'This woman's got five more hours of this, she'll just have to go a bit longer . . . she's selected a natural birth . . . if I give her nitrous oxide now . . . she's on the slippery slope to getting some other intervention' . . . so . . . so . . . like all I could do was really say what I thought I needed and I needed that and when they said, 'No' to that, I said, 'Well I *need* an *epidural*!' and that was when she actually went aside and spoke to my sister and my husband out of my hearing and they discussed what they thought . . . or what she thought . . . um . . . should happen . . . um . . . and ah . . . and that was when I . . . they put me in the bath . . . I didn't want to get in the bloody bath but you know it was like . . . I agreed because I thought anything might be better than what I was coping with (then hurriedly) what I was experiencing . . . um . . . but shortly after that she examined me and realised I was about to have the baby.
>
> (W, RHW Birth Centre client)

The theoretical dilemma brought out by this narration is not new. In championing that which has been marginalised or excluded by discourse there is always the danger of othering the supposedly dominant term. This risks remaining within an either/or logic where difference is massified and the opportunity for a diversity of practices is restricted. Is there any substantial difference between the paternalism of the obstetrician who confines the expectant mother in order to anaesthetise her and this midwife who restricted W's access to anaesthesia? In a sense, both are subsuming the specificity of the situation, standardising their practice to a pre-given schema which has as its rationale the noble motivation of acting in the mother's and baby's best interest.

W's text engages not only with the oppositions listed above but also with the discourse describing the cascade effect. W supposes that the midwife is thinking about this cascade effect as W's reference to the 'slippery slope' of intervention indicates. Receiving nitrous oxide is associated in this instance with the first step in an imagined sequence of events and interventions. What is apparent from this text is that the particularity of W's birth is subsumed into a project or pre-given schema. Implicit in this schema is a certain notion of time: that is, the time of project, or linear time. Inherent in this concept of time is a teleology[3] which W resists only by embracing the opposing teleology of medicalised birth. She seems fully aware that the call for an epidural is a sort of discursive trump card; that is, she threatens to upset the project of an unmedicated birth by crossing the oppositional divide between 'natural' and medicalised birth.

How does a space which is established to resist restrictive practices often associated with medicalised birth itself become the site for such practices? In the following section I consider the necessary role of language in the formation of identities which are, in turn, necessary for projects, in this case the political struggle for 'natural' or non-medicalised birth. It is suggested that the mobilisation of oppositions to create the space for alternative social practices may make for a certain rigidity, which risks the unwitting creation of other restrictive spaces.

The RHW birth centre as a site for the *chora*

Since the RHW Birth Centre appears to be an intermediate space constructed by, and between, strongly polarised identities, it is worth turning our attention to the issue of the formation of these identities. I have suggested that merely reversing hierarchically ordered oppositions need not escape the determinant ordering of space. In fact, there is a certain necessity to the ordering of space that coincides with our necessary ascription of identities to things through language. This is not to say, however, that the identities or spaces constructed through language are absolutely stable. Indeed, an understanding of the instability of these identities and spaces may enable less determinant imaginings and less rigid practices. Such an understanding, I suggest, must start with an exploration of the mutual constitution of language and subjectivity. To this end I think it is important to now turn to the work of Julia Kristeva, especially her work on the *chora*. For Kristeva (1984), it is a 'subject' who speaks but, rather than a subject who exists prior to language, subjectivity is formed in and through language.

Kristeva uses a psychoanalytic perspective to theorise the formation of the speaking subject. The simultaneous constitution of the subject and the subject's entry into the symbolic order arises at the moment that the infant separates from the mother. This is the moment when infants recognise that they are separate from their images and objects; ascribing symbols to their images and objects represents an attempt to overcome the distance of this separation (Kristeva 1984). To the extent that entry into the symbolic order entails the use of words, i.e. the capacity to signify and engage in the production of meaning, it is inseparable from participation in the social order (Kristeva 1981).

From a geographical perspective, Kristeva's use of a spatial notion to represent the period 'prior to' entry into the symbolic order is particularly significant.[4]

Kristeva uses the Platonic term *chora* to theorise an unordered, indeterminate space. Given that this unordered space is the pre-condition for symbolic language, it brings us closer to the implication of the material body in language, since it is the rule of bodily drives that governs an infant's realm of *signifiance* prior to separation from the mother.[5] Though the use of language implies a certain transcendence, Kristeva insists that our embodiment is never completely transcended, nor are its effects in language totally repressed.

Thus while Kristeva's use of a psychoanalytic perspective provides a developmental focus on the infant, her project is to theorise the continual *process* by which new identities form with each new utterance. This ongoing process arises from the tension of the dialectical relationship between 'symbolic language' and the material basis or drives which simultaneously disrupt *and* make language possible. Kristeva (1984) uses the term '*semiotic*' to describe this material aspect of the signifying process. While logically 'prior' to the use of language, insofar as the speaking subject is, for Kristeva (1984), a 'subject-in-process' (continually formed in and through the use of language) the semiotic pertains to the never-totally-repressed bodily drives. It manifests in the body, the voice, in writing and art as rhythm, timbre, repetition, pauses, laughter and so forth. Thus, where it is the symbolic that makes communication possible, the semiotic is inseparable from the creative dimension of the signifying process.

Kristeva's work offers an alternative theorisation of space to that which characterises oppositional politics. As the pre-condition for symbolic language, the space of the *chora* is a space of indeterminacy, an alternative to the ordered space of the symbolic and its hierarchically opposed identities. The *semiotic*, as the material aspect of signification, has the potential to disrupt or destabilise the symbolic and thus the social order. Consider the semiotic in the following transcript.

> It really wasn't until we were there for about half an hour that I . . . that it sort of hit me that she was . . . like she was in serious *serious* labour (laughs) probably the most . . . the most telling fact . . . um . . . was when I was behind C and holding her through contractions and it was the muscles . . . the contractions through her muscles that made me think that this is just . . . this is an *amazing* thing, because the muscles were contracting so hard that no matter how much pressure I put on to try and relieve . . . because she sort of wanted me to . . . I had my hands around the back . . . sort of around the kidney area and it was like a rock and that . . . that's when it hit me . . . it was about then which was about ten o'clock that night in the . . . ah . . . Birth Centre.
>
> (S, RHW Birth Centre client)

In the same way that we can never truly represent the semiotic we cannot say that the birth *centre is* the space of the *chora*. However, we can situate the *chora* in the birth centre; that is, we can lend it a topology. Situating the *chora* at the birth centre allows the body to enter into geography in a way that merely writing about the body as an object cannot. The repetition of the words 'muscles' and 'contractions' intimate the very repetitive materiality of the contractions themselves. The laughter after 'serious *serious*' and the emphasis on the word 'amazing' may be seen to indicate something of the inexpressible, but not totally repressible elements, as S attempts to put this very bodily experience into language. Thus, in focusing on the materiality of a narration such as S's, that is, on the semiotic, the body speaks.

By situating the *chora* in the birth centre, indeterminacy, which seems particularly important in an experience like childbirth, can be theorised. Where the dualist thought presented in Table 6.1 is associated with an ordering of experience (evident in W's perception of the slippery slope to intervention) the conceptualisation of indeterminacy enables an appreciation of a multiplicity of experiences. For example, an openness to the particularity of experiences also entails an ability to appreciate other, non-linear ways of thinking of time. In terms of the political project of 'natural childbirth', an experience such as S's may be of limited use. However, in relation to other concepts of time, S's experience takes on far more importance: here I am referring particularly to what Kristeva (1981) calls cyclical time. It comes through most clearly in the repetition in the telling of the experience itself, the cyclical experience of the movement of time echoing through the evocation of the cycles of muscles contracting.

To be able to think of birth outside linear time (the time of project) and outside a narrow political construction is extremely important. This is not to deny the importance of linear time; as J shows in her appreciation of the intimacy of the birth centre space, the dualisms associated with the time of project can open up space. However, a failure to take account of other instances of time may represent a closure to the particularity of the present moment. Importantly, Kristeva's (1981) 'Women's Time' calls for the interweaving of different times and the recognition of the co-existence of contradictory ways of appreciating time. Consider the difficulty C has in this final transcript subsuming her experience into the linear time of project:

> Time is just . . . is irrelevant . . . you know you sort of think . . . you know over the years I've heard time is a man-made concept and you know you ponder on that . . . but when you're birthing it's just . . . that is just so real it's unbelievable...

and, when pressed for ways of describing time,

> Time is just totally irrelevant . . . it's just like . . . um . . . incidences in relation to your body and what you're trying to do with it . . . I guess it's over a space of time but . . . but if that whole experience had happened over four days or you know had happened over three hours as it kind of did in that birth centre I don't think I would've known the difference you know and that seems like a really odd thing to say . . . but I wouldn't have had any idea of the time . . . it's the strangest thing . . . and it's one of the nicest experiences I've had in that way too because you're totally focused on something . . . important and very real I 'spose . . . time and space are irrelevant as long as you feel comfortable.
>
> (C, RHW Birth Centre client)

The difficulty C has in describing time, the materiality of the narration, with its pauses and repetition in the claim that 'time is irrelevant' and the materiality of the experience hinted in the phrase 'incidences in your body and what you're trying to do with it' are, at best, intimations of the *chora*, the space of the semiotic. C's experience can be seen as an instance of the cyclical time of repetition which Kristeva (1981: 16) associates with 'extra-subjective time' and 'unnameable *jouissance*'. *Jouissance*, which defies any simple translation but means something like blissful, ecstatic or self-erasing joy, may be detected by the words 'one of the strangest and one of the nicest experiences' which C attributes to the focus solely on labour: 'something very important and real'.

Conclusion

The purpose of this chapter has certainly not been to argue that we should or can do away with the dualisms that have constructed contemporary debates around childbirth. In fact, I have demonstrated their importance in opening up a space for alternative practice, while arguing that our thinking need not be completely constrained by such dualist thought. A worthwhile project for geography, then, and one that can meaningfully reincorporate the corporeal, is to give voice to the semiotic, necessarily through the symbolic. Such a project starts from the recognition that the necessity of bringing the semiotic into the symbolic entails a process of excising a 'thesis' from a rich and unordered dimension of social life which is yet to be inscribed with meaning (Kristeva 1984). Where Kristeva's work is important is in her recognition that this entails a ritualised violence, in that we must carve what becomes provisionally determinant out of indeterminate flux. Such an awareness allows a minimisation of this violence, capturing it where it is most embryonic: at the moment of identity formation. For the practices of childbirth, a focus on the formation of identities may enable a demassification of differences. Where the strong polarisation of the debate into medicalised and non-medicalised birth has led to a certain restriction of practices, demassifying these poles may allow for a greater diversity of experiences to count for a 'significant' or 'successful' birth. The situation of the *chora* in the RHW Birth Centre allows an attentiveness to the indeterminacy and opportunity of such a space, and to different notions of time. The value of this attentiveness is that the haste to fulfil future projects is not bought at the too-high-cost of stumbling clumsily over the significance of the present.

Acknowledgements

Thanks particularly to Maria Hynes and Bob Fagan for reading earlier drafts of this paper.

Notes

1 While the split between mind and body has been at the centre of much intellectual writing of this century, it has often been assumed that conceptualising the corporeal, or 'bringing the body back into theory', will overcome the mind/body split. Challenging this assumption, Kristeva's work recognises that the very notion of writing *about* the body indicates a separation inherent in symbolic language. Indeed, anatomy and physiology are disciplines that conceptualise the corporeal while maintaining a clear distinction between mind and body. Kristeva's work, in contrast, enables a consideration of this distinction as, in a certain sense necessary, whilst in no ways given or stable.

2 This unstructured model does not imply that the researcher (despite being open to the directions in which the informant's conversation might lead) is or can be a 'blank slate'. For this research, prompt words pertaining to the themes of ordering, the body, space, social relations and practices guided the interview. Tape-recording of the interviews allowed concentration on the actual conversation itself, while of course providing a record for close analysis of what was being said and, crucially for this research, *how* it was being said.

3 Teleology assumes a forward movement towards an end as a seemingly natural progression, and thus comprises an implicitly linear concept of time.

4 This use of 'prior to' cannot be understood in any chronological sense since the ordering of chronological time itself belongs to the symbolic realm.

5 Kristeva uses the term *signifiance* to register that the body is neither entirely of the order of signification (the symbolic order), nor completely excluded from the signifying process. Kristeva (1984: 17) writes: '(w)hat we call *signifiance*, then, is precisely this unlimited and unbounded generating process, this unceasing operation of the drives toward, in and through language'.

References

Arms, S. (1975) *Immaculate Deception: A New Look at Women and Childbirth*, New York: Bantam Books.

Balaskas, J. (1983) *Active Birth*, London: Unwin.

Clarke, J. (1996) 'Happy Birthday', in D. Adelaide (ed.) *Mother Love: Stories about Births, Babies and Beyond*, Sydney: Random House, 251–271.

Gibson-Graham, J.K. (1997) 'Postmodern becomings: from rape space to pregnant space', in G.B. Benko and U. Strohmayer (eds) *Space and Social Theory: Geographical Interpretations of Postmodernity*, Oxford: Blackwell, 306–323.

Harkness, L. (1986) *Birth: Where and How*, Surry Hills, NSW, Australia: Horan Wall and Walker.

Information about our Birth Centre (1993) Paddington, Sydney: Royal Hospital for Women.

Kristeva, J. (1981) 'Women's time', *Signs: Journal of Women in Culture and Society*, 7, 1: 1–35.

—— (1984) *Revolution in Poetic Language*, trans. M. Waller, New York: Columbia University Press.

Marcus, S. (1992) 'Fighting bodies, fighting words: a theory and politics of rape prevention', in J. Butler and J. Scott (eds) *Feminists Theorize the Political*, London: Routledge, 385–403.

Martin, E. (1990) *The Woman in the Body: A Cultural Analysis of Reproduction*, Boston: Beacon Press.

Minichiello, V., Aroni, R., Timewell, E. and Alexander, L. (1995) *In-depth Interviewing: Principles, Techniques, Analysis*, Melbourne: Longman.

Oakley, A. (1992) *Social Support and Motherhood: The Natural History of a Research Project*, Oxford: Blackwell.

Philosophy of the RHW Birth Centre (1992) (revised) Paddington, Sydney: Royal Hospital for Women.

Positive Choice (undated) Paddington, Sydney: Birth Centre, Royal Hospital for Women.

Pringle, R. (1998) *Sex and Medicine: Gender, Power and Authority in the Medical Profession*, Cambridge and New York: Cambridge University Press.

RHW Birth Centre Protocol: Client's Copy (undated) Paddington, Sydney: Royal Hospital for Women.

Rose, G. (1993) *Feminism and Geography: The Limits of Geographical Knowledge*, Cambridge: Polity Press.

Saul, A. (1994) Theoretical and political implications of the management of childbirth, paper presented to *Issues of Contemporary Sociology Seminar*, University of Newcastle, September.

—— (1995) 'It'll all come out in the meta-analysis: debates about feminist methodology and the politics of the management of childbirth', paper presented to the Second Newcastle Interdisciplinary Gender Studies Conference, University of Newcastle, June.

Taylor, R. (1979) *Medicine Out of Control: An Anatomy of Malignant Technology*, Melbourne: Sun Books.

Whelan, A. (1994) *Centering Birth: A Prospective Cohort Study of Birth Centres and Labour Wards*, unpublished Ph.D. thesis, University of Sydney.

7

PUTTING PARENTS IN THEIR PLACE

Child-rearing rites and gender politics

Stuart C. Aitken (USA)

> I saw that we were about to embark on a familiar set of steps in an ancient dance. She was to become more and more the unprotected female in the presence of gigantic male creatures . . . We, for our part, were to advance with a threatening good humour so that in terror she would have to throw herself on our mercy, appeal to our generosity, appeal to our chivalry perhaps: and all the time the animal spirits, the . . . 'amorous propensities' of both sexes would be excited to that state, that *ambiance*, in which such creatures as she is or has been, have their being. This was a distancing thought and brought me to something else. The size, the scale, was wrong.
>
> (William Golding 1980: 57–58)

> There seems some reason to believe that the male imagination, undisciplined and uninformed by immediate bodily clues or immediate bodily experience, may have contributed disproportionately to the cultural superstructure of belief and practice regarding child-bearing.
>
> (Margaret Mead 1955: 222)

Many view rites of passage quite positively. They argue that rites constitute important gateways in the stages of the life cycle. Through a set of culturally prescribed codes and mores, initiates may be honoured by family, community and society, and gain support and guidance from those who have gone before.

Others interpret rites of passage somewhat differently. They suggest that rites and rituals define sources of strength and empowerment on a journey to spiritual wholeness; and some are concerned about the lack of rites in contemporary Western society. Mythologist, Joseph Campbell (1949, 1972) and cultural critic,

Ronald Grimes (1995), for example, are concerned that contemporary society may well suffer when it no longer sanctions specific rites of passage. Campbell (1949: 11) argues that rites are effective spiritual aids which supply the symbols that carry the human spirit forward, and he feels that, without them, the foundations of society crumble. Michael Meade (1996: 27) suggests that the lack of rites is associated with a form of mass denial that may relate to increases in random violence and drug abuse, to the collapse of medicine, and to the confusion in personal, gender and national identities. These are heavy charges that suggest the importance of rites and rituals in many parts of our lives – but they are charges that leave unquestioned the social order and scale from which rites derive their power.

With this chapter, I want to caution against the universal appeal of child-rearing as a rite of passage. I argue that there are certain covert (concealed) cultural norms associated with child-rearing in Western society that enervate rather than empower. I use the term 'enervate' to mean 'to curtail the power of the agent'. Thus, we need to reconsider carefully the contemporary wisdom of rites as they are applied to new parents. Covert cultural norms and sex-gender meanings can be detected through a detailed understanding of the complex interplay between an ideal sex-gender concept, actual behaviours and the strategic use of sex-gender performances (Del Castillo 1993). I argue that, despite an apparent slackening of reproductive rituals and beliefs over the last several decades, rites are translated at a larger scale into covert cultural norms that constrain the activities and responsibilities of women and, to a lesser extent, men. I am persuaded by Roland Barthes' (1972: 155) suggestion that rites of passage are a form of myth that, in producing covert cultural norms, immobilises rather than enables. Barthes argues that rites suggest and mimic a hierarchical order which, at the very least, stultifies creativity and intuition, but also clearly indicates, or marks, the bearers of power as well as those who are subservient.

The questions I raise relate to the wholeness of the wider context into which young parents are thrust by the rites associated with child-rearing. Put simply, what are the new social identities that new fathers and mothers embrace with parenthood? And in what sense is the work of parenting enervating? The focus in this chapter is on how the notions of motherhood and fatherhood contextualise, but also constrain, parents' attempts to define themselves and their place in the world.

A longitudinal research study of young families in San Diego, California has enabled me to explore changes from prior to the birth of a first child to until the child is 4 years old. The participants were derived from a large survey sample of women who visited one of eight local obstetrical-gynaecological clinics when first pregnant. From this sample, we contacted women who were expecting a first child. One hundred and twenty-seven in-depth interviews were conducted with adults from households of the pregnant women. Over half of these initial participants remained with us through a series of yearly interviews.[1] In the larger study (Aitken 1998) I argue at length that the institutions of motherhood and fatherhood are grounded in a mythic family structure that is no longer workable. In this chapter, I focus on day-to-day parenting, unravelling it to reveal it as a critical constructor of identity that is, at the same time, obscured by covert cultural norms.

The rite to patriarchy

The chapter's opening epigraph by William Golding hints at issues of scale that are rarely considered in the literature on rites of passage. When originally conceptualising 'rites of passage' in his highly influential thesis, Arnold van Gennep (1960) argued that a triad of separation, transition and incorporation could be identified in most ritual observations. Amongst some feminist anthropologists (e.g. Homans 1994; Kitzinger 1994), the transformation to patriarchy is theorised as specific transition-rites (e.g. birthing rituals) that are symbolic and focused within a narrow time-scale and contextualised locally (i.e. seen in their local time–space settings). I want to argue that this is particularly problematic because it ignores the scale of patriarchy and the complex relations between the myth of parenthood and the reification[2] of patriarchy in day-to-day practices of child-rearing over an extended period of time. Motherhood and fatherhood, although often appearing 'natural' because they are implicated with the continuous practice of mothering and fathering, in actuality are problematic social constructions because they derive from larger scale rites and rituals (Figure 7.1).[3]

There is a fairly substantial and fascinating literature on the specifics of childbirth as a rite of passage from one social order into another. In Western society childbirth rites are varied and diverse but may include Lamaze[4] classes, dietary rules, baby showers, building cribs, painting rooms, sexual taboos, naming ceremonies, the handing out of cigars, baptism and circumcision. Some recent studies focus on how birthing rites are often professionalised within scientifically based institutionalised obstetrics and gynaecology (MacCormack 1994; Kahn 1995). These studies uniformly agree that birthing rituals operate within specific social contexts and value systems. They acknowledge that these rituals play a role in reproducing values held by politically dominant groups and encourage a particular sexual and social division of labour and a larger patriarchal order.

This chapter is concerned with what happens after a new child and mother return to the home. The day-to-day work of parenting is complex and intractable to the extent that important issues of inequality and dissent as well as consent, contest and contradiction are lost in an imprecise and convoluted coalescing of gender roles and relations. Simultaneously, at a scale of resolution that ignores

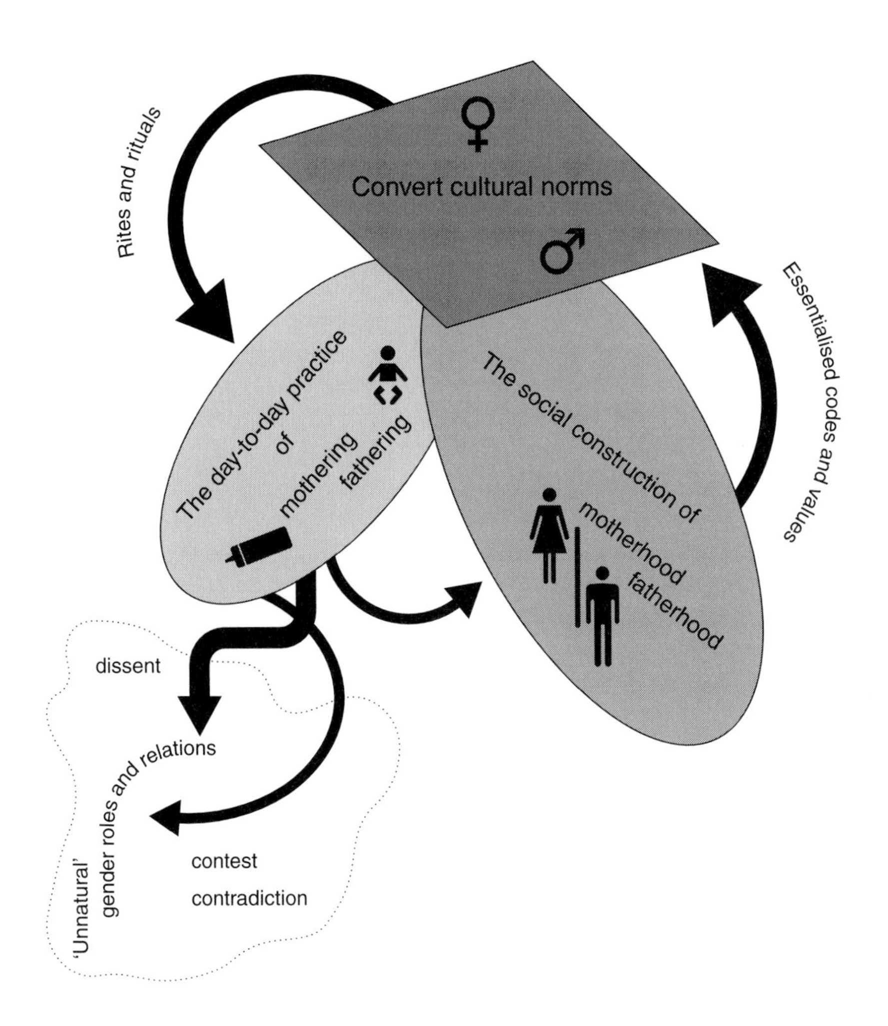

Figure 7.1 The scale of parenting

the local, the body and the day-to-day, codes and mores are derived from a seemingly unproblematic acceptance of motherhood and fatherhood as natural categories. These categories are translated into covert cultural norms that are unworkable for contemporary parents because they reflect only partially the work of mothering and fathering.

The chapter's second epigraph by Margaret Mead notes the importance of understanding child-bearing at the scale of the body. Our understanding of the rites of child-bearing and child-rearing rituals neglects the scale of the body, the everyday and the local, and is steeped in the patriarchal codes of elder and dominant social groups and, for the most part, our understanding of these. For example, in the tradition established by van Gennep, transition-rite theorists embrace grand dichotomies that artificially separate the sacred and the profane, death and re-birth, young and old, male and female; they do so at a societal scale that refracts rather than reflects locales and bodies; and several of them would have us believe that it is solely with abstract rites, rituals, codes and mores that identities are imagined and represented to us, rather than in the context of the everyday. Mead's concern was that it is a dominant male imagination that contributes most to the culture and practice of child-bearing, yet it fails to understand the often unpredictable and sometimes irrational penchant of bodily experiences. I would argue further that, in this context, the male imagination is constrained to a scale that does not recognise the importance of the body and the local. How we imagine child-rearing should not be bound by scale, but should circulate freely with unlimited social mobility that includes the body and the local as well as the societal.

Transition-rite theorists rarely articulate this complexity. They suggest that, as an image and an event, a rite-transition reinforces culturally prescribed age roles and sex roles by dramatising individuals' transition to a new role and educating all members of the community about that role's rights and obligations (Paige and Paige 1981: 3). The role change is simple and straightforward compared to the extraordinary complexity of the everyday. Transition-rite theorist, Frank Young (1965: 111–112), for example, unproblematically interprets the birth process as a rite of passage for both parents that focuses upon parenthood as it relates to 'solidarity' to some larger family unit with implicit sexual roles and relations. Thus, he sees motherhood and fatherhood as performances made meaningful only in relation to larger covert cultural norms; the context of the body, the everyday and the local must be subsumed within this larger polity. But, at the day-to-day level, the complex work of parenting is not sufficiently understood to allow such simple assertions. The simplistic assumption that the nature of motherhood and fatherhood is unproblematic casts 'unnatural' parenting practices (by single parents, same-sex parents, househusbands and so forth) into the contested arena of the marginal and socially ostracised.

I propose an alternative argument, i.e. that there is nothing 'natural' about child-rearing. The complexity of child-rearing stimulates the kinds of changes that highlight the structural fragility of contemporary families. Covert cultural norms are nothing more than misplaced attempts to bind the family as a functioning unit. At the day-to-day level there is a complexity that embraces negotiation and resistance, but, at some larger scale, that negotiation breaks down into a reified and gendered division of labour that is patriarchal.

Part of my incentive to embark upon the long-term San Diego study was to seek out men and women who were breaking the traditional gender roles of fatherhood and motherhood. We learnt that although some men were taking on more domestic and child-rearing roles and many women were engaging the waged world of employment, patriarchal gender relations, for the most part, remained, and long-term child-rearing responsibilities most often befell women. What follows are vignettes from some of the child-rearing stories we heard. The vignettes are selected to illustrate specific points and are not intended to be replicable or generalisable. In the interviews we did not focus specifically on rites, rituals and larger issues of identity and, as a consequence, the fieldwork limits how I can write about these larger processes. None the less, contexts of political and cultural identity, patriarchy and sexism are thinly veiled in many of the vignettes. In what follows, I broach a limited set of themes relating to transition-rites. The themes are broadly arranged to reflect the influence of elders, parenthood and the space–time work of parenting. They are introduced by first using data from the larger survey sample of first-time parents; this is followed by more detailed discussion of the changing life spaces described in the stories of three specific couples.

The influence and space of elders

Transition-rites theorists highlight the influence of elders in the maintenance of particular social orders. Hilary Homans (1994: 221) notes that pregnancy, birth and child-rearing rituals have a particular role in reproducing values held by elder groups, encouraging a particular sexual and social division of labour. Based upon our large mail-in survey of households expecting a new child, Figure 7.2a suggests that *prior to the birth*, first-time parents perceive that their parents will be the greatest source of guidance on raising their new child but *by the time the child is 1 year old*, the influence of parents decreases slightly and the influence of friends with young children increases substantially (Figure 7.2b). Conversely, over the same period, young parents agree that their own parents are an important example to follow and this perspective is maintained after the birth of the child (Figure 7.3). In-depth interviews allow the aggregate statistics to be probed.

Anarosela and Michael are an interesting example of elder influence because, like many low-income young couples expecting a first child in San Diego, they reside with Michael's parents. Through the course of our interviews, the relations between the two generations shifted in three significant ways. First, as the pressure of his employment increased, Michael's role as a father-figure grew although – ironically – his work at parenting lessened; second, Anarosela's identity as a mother changed in unanticipated ways with Michael's increasing comfort with fatherhood; third, Michael's status in the household was elevated after his father's debilitating stroke. Underlying some of these tensions is the

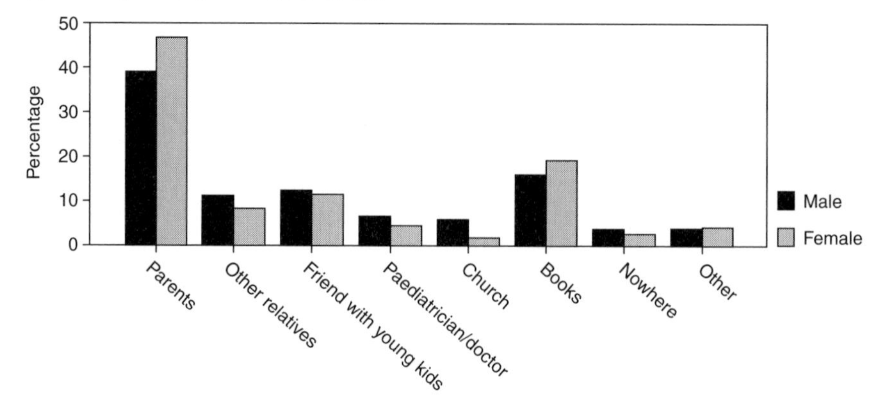

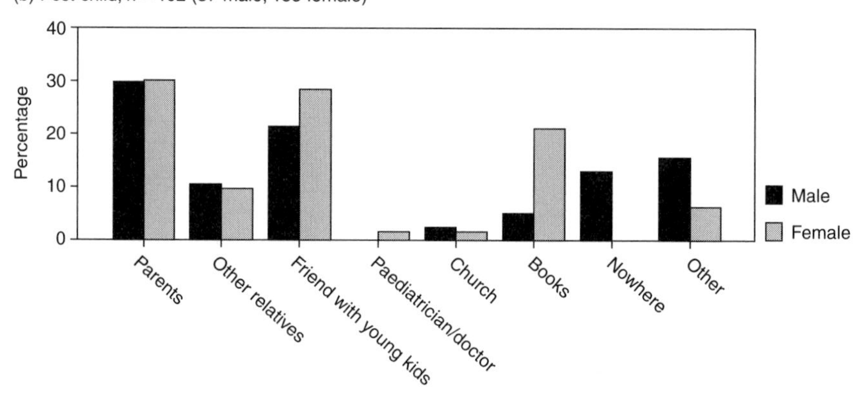

Figure 7.2 Sources of guidance for first-time parents (a) perceptions prior to birth; (b) perceptions one year after birth

presence of Michael's mother, her influence on Anarosela's domestic work and her diminished attention towards her grand-daughter after her husband's stroke.

We first contacted Anarosela and Michael in early 1992 when Anarosela was pregnant with their first child. Anarosela was born in Tijuana, Mexico, but, aged 17, moved north to San Diego. We met Anarosela three years later, pregnant, and just having quit her job. At that time, Anarosela noted that she wanted to go back to work right after their child was born. Due primarily to a lack of financial resources, the couple moved in with Michael's parents. After Simone was born, Anarosela said that she felt some pressure from her mother-in-law to conform to what constitutes appropriate motherly behaviour, but she was careful during interviews not to be too critical.

Statement: Your parents will be an important example for you to follow when raising your child
(a) Pre-child; $n = 259$ (104 male; 155 female)

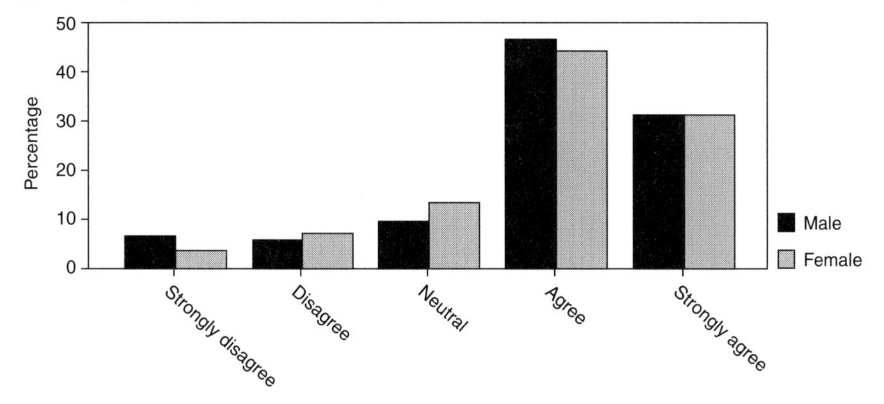

Statement: Your parents are an important example for you to follow when raising your child.
(b) Post-child; $n = 97$ (37 male; 60 female)

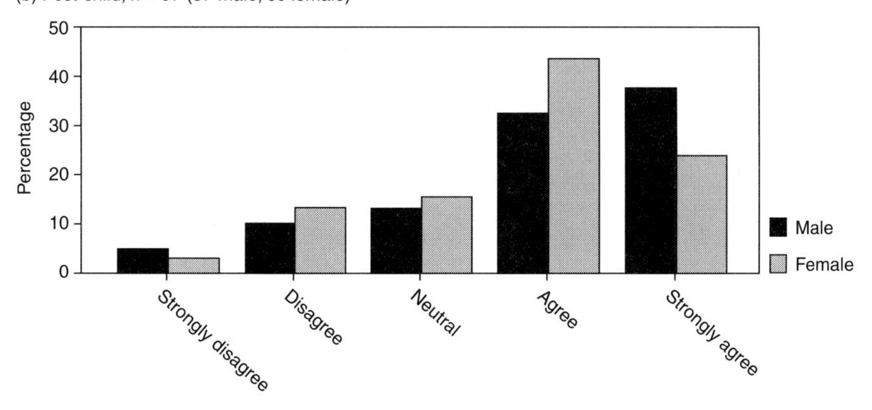

Figure 7.3 Importance of elder parents as example (a) perceptions prior to birth; (b) perceptions one year after birth

Michael's parents live in a spacious 1960s bungalow on Otay Mesa, within a few miles of the Mexican border. When they first moved in, Anarosela and Michael really liked the suburban feel of the neighbourhood. Their feelings of well-being dissipated a little as they noticed gang activities in the area. Michael put it this way:

> We've got our problems with gangs sometimes. In the back because we've got those windows upstairs we see things going on. There's always something going on, that's the bad part of this area . . . I guess nowadays it is even worse . . . We know [some of] the neighbours. We don't have a relationship with the ones in front of us because they are hardly ever there and we don't know anybody from that house.

Anarosela pointed out that she would not walk in the neighbourhood by herself during the day or at night because of the gang activities. Her fears came into even sharper relief when Simone was born. Several women in our study noted a heightened fear of public spaces after their children were born which may be part of a general apprehension over the well-being of children in some contemporary American communities. Without overt community connections or secure feelings in local public spaces, isolated parents may withdraw into what they feel is a secure, safe home environment (Aitken 1994). In addition, for young, low-income couples with few choices, a parent's home is often encountered not only as a legacy, but also a set of rules and practices.

Simone was born in May 1992 and by the time of our interviews in June, Anarosela was taking care of many of the household chores as well as the baby's needs. By this interview, she had decided not to go back to work for a while and stated that she thought it 'bad when mothers only take a month off to have their children'. This statement oddly contradicted her sentiment at the first interview and we wondered at the influence of her mother-in-law. At this time, Anarosela was not going out at all and asserted that she had no free-time for herself although she tried to arrange things so that Michael could go out some evenings after work. She admitted to being over-protective of Simone who was born with spina bifida. When Michael offered to help, she said she often shrugged him off as being unsuited to give the kind of care the baby requires. At this interview, Anarosela rolled her eyes and made a comical face when asked about whether her mother-in-law's help eased her parenting.

A year later, during our interviews in September 1993, Anarosela was more vocal about the tension with her mother-in-law. Both she and Michael spoke of increased resentment. Anarosela was particularly upset that nearly all the household chores had now become her responsibility:

> My father-in-law does most of the shopping for the household and [my mother-in-law] helps a lot with the looking after of Simone, especially putting her to sleep. But she does no cooking or house-cleaning.

Michael was more pointed about his mother's contribution to Simone's care, saying that he felt she resented the burden and probably would prefer not to do any work with the child. Anarosela's Hispanic background, he noted, stopped her from saying anything too disrespectful about his mother. Tension in the family was heightened at this time by the arrival of Michael's sister. The sister's stay was only for two months because 'things did not work'. As a parting gift, Michael's dad gave the sister one of the household cars causing significant tension because Michael and Anarosela were now responsible for getting Michael's mother to her numerous appointments with doctors.

Simone was 14 months at the September 1993 interviews and Anarosela had not resumed any form of paid employment. She could see no way of attaining her wish of getting back to school in the next two to three years. A second significant source of frustration for Anarosela was her 'motherly' relations with Simone and her perception of Michael's role as a father. Anarosela's highly protective mothering, evident in the first set of interviews, contextualised a large part of her relationship with her daughter. Michael's fathering, on the other hand, changed significantly between interviews. Greater job security due to his length of tenure at the restaurant enabled him to focus more attention on his daughter. Although his job required that he spend less time with Simone, Michael's lack of stress about work enabled him to have quality time with his daughter which resulted in confident fatherhood. During our interview with Anarosela, Michael spent the whole time playing with Simone in the adjoining room. In the previous interview, his interaction with his daughter was much less intimate.

Our last set of interviews with Michael and Anarosela were in March 1996, just prior to Simone's fourth birthday. We did not interview the couple in 1995 because Michael's father was recovering from a severe stroke. Anarosela pointed out that her father-in-law needed as much care as Simone, and her mother-in-law was now unable to do anything with her grand-daughter or around the house. Now, all domestic chores fell to Michael and Anarosela. Pregnant once more, Anarosela was resigned to her receding dream of going back to school:

> I guess if I really wanted to go back to college and finish my general classes I would. But I have to take [Simone] a lot of places like her speech therapy, her physical therapy and now her school and sometimes it's hard to get to. But probably if I really try hard I could do it. It's something that I really want to do. And with another one coming along the way it's going to be really hard.

After his father's stroke, Michael's role in the family changed to that of 'head of household'. Michael took this new position seriously and envisaged their future as consisting of the extended family. Anarosela had this to say about changing roles in the household:

> I would say I have fifty-fifty control over raising Simone, but when it is like something regarding the house I can stay out of it and let my husband and my mother-in-law decide. When it is between them, I just stay out of it.

Anarosela noted that money was tight and their finances could be better, but she felt less stress than she had at previous interviews. She and Michael had a close friendship and seemed content with their circumstances. Neither suggested any intentions of leaving the household. Michael was now the 'head' and Anarosela was beginning to carve her own matriarchal identity as the presence of her mother-in-law receded towards the care of her ailing husband. It was clear that power relations within the household were subtly transforming. It seemed also that Anarosela was developing some jealousy around Michael's bonding with Simone:

> Well, I feel like I'm a good parent but I know I could be better. When I am with her, my daughter, I know she respect me as much as she respect my husband. [But] if I tell her to come here then I have to tell her 'Well, I'm going to count to three and if you don't come . . . ' And I have to tell her three or four time. She listens to me when he's around. When he's around she does perfectly. When he's working I have to tell her and tell her and tell her, and finally when she gets – I have to go get her and I say, 'I'm going to take you by the ear', and that's when she say, 'okay Mummy I'm coming, I'm coming'. But she gives me attitude. She's not even 4 and she's not that bad. To her, her Daddy is the King and I'm the Witch. She'll see when she's 18 and she wants to go out dancing; who's going to say no!

This brief vignette suggests an interesting interplay between the domestic and parenting work that Michael and Anarosela took on, and changing relations with elders that constituted a re-positioning of family power. The couple are relatively content with their parenthood and relationship. Michael's gradual assumption of fatherhood and his dramatic assent to the role of patriarch suggests an emerging political identity with which he is comfortable. Anarosela's resentments are highlighted in her changing relations with her daughter and the increasing distancing of her dream of going back to school. The vignette may suggest an upwelling of patriarchal codes and mores as a young couple take on their elders' rites and responsibilities but there is clearly more to the work of their parenting than this brief account suggests. Although their lives seem to bear its marks, neither Michael nor Anarosela think in terms of patriarchy as a constraint nor would they consider the ways power relations are embodied within the household. Nonetheless, the subtle changes evident from one interview to the next suggest a complex interplay between covert cultural norms and day-to-day parenting. A similar interweaving is suggested in the next two vignettes where there are no elders present on a day-to-day basis, and household space and domestic responsibilities are shared equally.

Parenthood, employment and the work of parenting

Figure 7.4a suggests that, for the larger sample, there is a strong perception that a new child will disrupt the time afforded discretionary activities. Agreement with this sentiment increases by about 10 per cent once the child is born (Figure 7.4b). Many young parents agree that part of their rites and responsibilities is to lose control of their time because of the demands of their children. Two new mothers put it this way (quoted in Aitken 1998: 84):

> My schedule has always lived for me, and now I really live around [my son's] schedule . . . I think so far I've realised that if I live by his schedule, we all get along a lot better.

> When I really like to come home and relax after work, I have to hurry up . . . So, it's like I have to make every minute count for something. There's really not a lot of idle time to just piss around.

Figure 7.5 reveals an interesting bipolarity in the way men and women perceive a new child disrupting their paid employment. For the most part, men are less likely than women to perceive that a new child will disrupt their employment, some disagreeing emphatically with this sentiment. After the child is born, fewer men disagree that the child will disrupt their employment (Figure 7.5b). Women, on the other hand, are clearly more realistic about the impact of a new child on their paid employment.

Lisa and John believe that they share work and time in an equitable way; a belief that continued through the four years that we knew them. Lisa is a baker and John – specialised in tool and die-casting – had just lost his employment in a machine shop a month prior to our first set of interviews. At that time, John was comfortable that they had saved enough money to ride out the early 1990s economic downturn in California. It was clear that the couple were comfortable with their domestic gender roles:

> We do food shopping. That's kind of like the thing we do together. That's important. I do my own laundry every week, in fact I'm doing it right now (cocks ear to the noise of the washing machine). I don't do any food preparation; I'm a lousy cook! But we split the chores up: I do the house cleaning – all the vacuuming and stuff and she does the food; I do dishes too but I don't do them as often as she does.

Lisa was pre-occupied with her father's recent hospitalisation and had less to say at this first interview. She noted that she would get less sleep once the baby was born, but also that she wanted to continue to work outside of the home. She hoped that the new child would bring her and John closer together. Regarding the equity in household chores, she noted that John helped out in many things when she was pressed for time, and that cleaning was exclusively his responsibility.

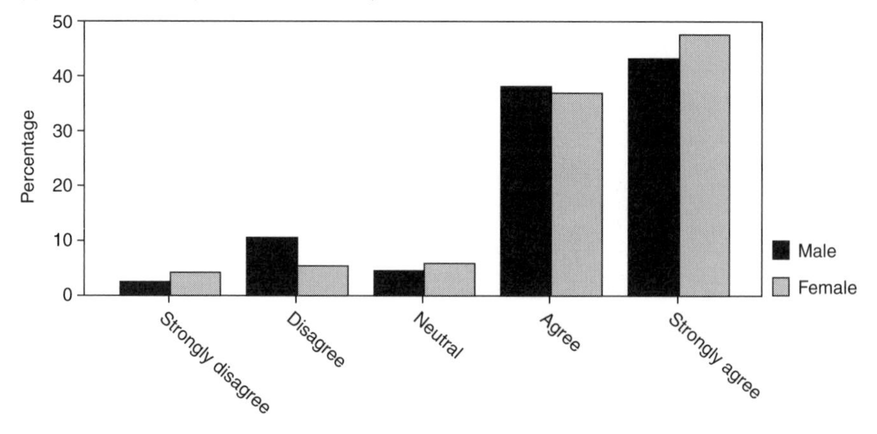

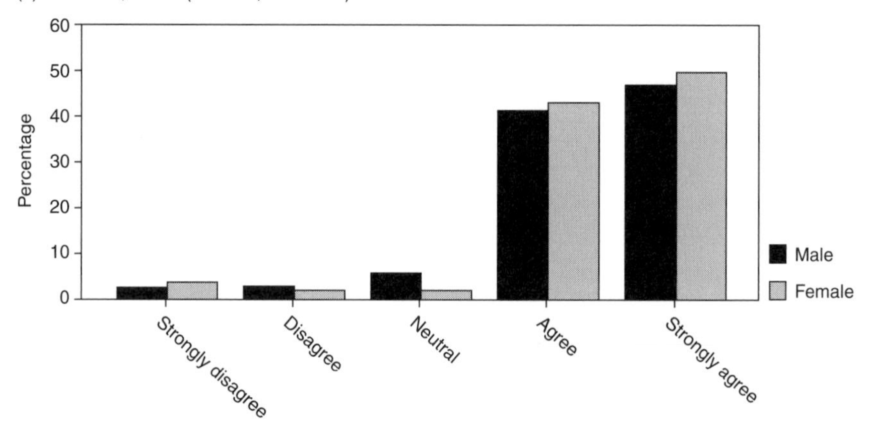

Figure 7.4 Perceptions of disruptions to free time (a) prior to birth; (b) perceptions one year after birth

Both were optimistic about parenthood. John said that he was looking forward to sharing all the good and bad things of this world with his child. He noted that he was ready to inherit the 'long awaited role as a father' and was in agreement, with some cryptic reservations, that his parents would be an important example to follow when he raised his new child. When asked if their new child would change their lives John stated that:

> I don't have fantasies about looking forward to childbirth: 'Oh yeah, everything's going to be perfect'. It ain't that way! Going through Lamaze

Statement: The new child will disrupt your employment
(a) Pre-child; $n = 259$ (104 male; 155 female)

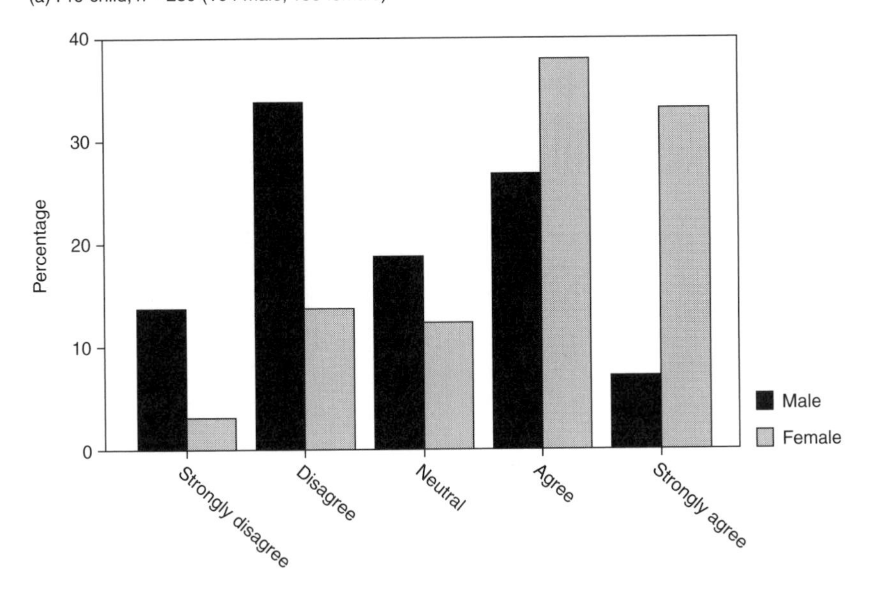

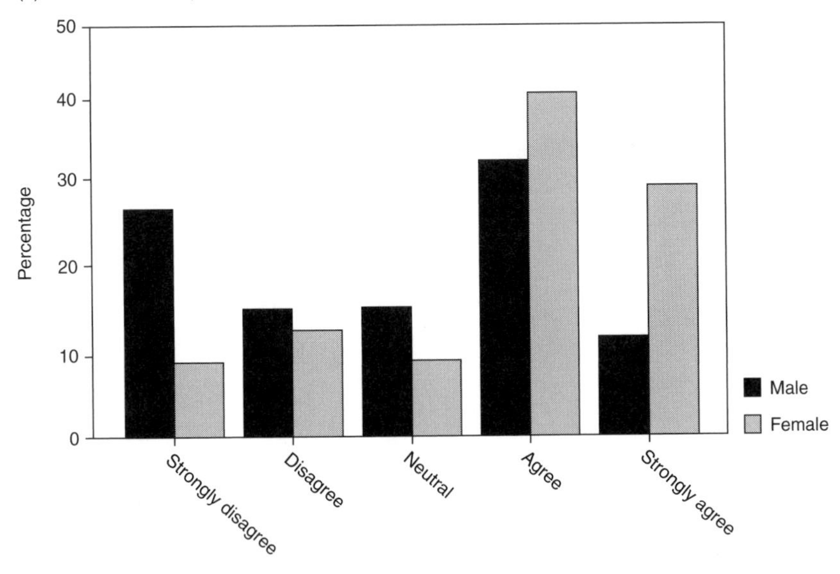

Figure 7.5 Perceptions of disruptions to employment: (a) prior to birth; (b) perceptions one year after birth

> classes we learnt a lot, and we know pretty realistically what to expect, I think (laughs) we are prepared as we can be for childbirth. I guess you never really [are] totally. We're going to learn a lot; maybe this is going to change, I don't know. My parents? Yeah, unfortunately in some respects they will influence me (laughs). Okay, the world's still wonderful and we're going to bring the kid up to really realise the beauty in it. He'll learn all the bad things as he goes along. That's what we're going to try to emphasise anyway: just positive images and positive things for him.

John was particularly circumspect about how the new child might limit his job potential. He opined that it would help develop in him tools – such as patience – that would be a positive benefit in his field.

When we interviewed John and Lisa a year later, John had gained temporary employment as a bread sales representative. He got up every morning at 2 a.m. to make deliveries in addition to going to school three nights a week. The rest of his waking hours he tried to spend in quality time with Heather. Table 7.1 is a comparative time budget representation of John's and Lisa's day before the interview prior to Heather's birth and then a year later.

> I don't get to spend as much time with her as I'd like. I miss her when I don't get to see her. Actually, I've been spending some quality time with her during the day, but some days I only get to spend, like half-an-hour. When she was first born I was laid-off so for the first six months she didn't have child-care: I took care of her. My wife worked so I played Mr Mom.

John's sentiments join with those of other 'house-husbands' who participated in our study (cf. Aitken 1998: 62–65). Contemporary Western rites of fatherhood do not include full-time child-tending and, when engaged in those activities, our male participants 'played' at being mother. Sara Ruddick (1992) envisages all parenting of this kind as mothering. She argues that fatherhood, as a concept, detaches emotion from the work of raising a child. This detachment, she claims, leaves fathers with love for their children but little responsibility for the work that such emotions entail. Child-tending men are disembodied father-figures because they cannot take on the work and responsibility of parenting without imagining themselves as 'Mr Mom'. In a bold attempt to re-imagine the essentialist rites of motherhood and fatherhood, Ruddick suggests 'mothering' rather than 'parenting' as a gender-inclusive and therefore genderless descriptor of the work in which child-tending men and women engage.

John had an intimate bond with his daughter that was evident as he played with her during the interview. He also took care of Heather's needs as they

Table 7.1 Time budget representation of John and Lisa's day before interview

	Pre-child		*One year after Heather was born*	
	John	*Lisa*	*John*	*Lisa*
Midnight				
1 a.m.				
2 a.m.			woke and got ready	
3 a.m.			driver for bakery ↓	woke up and dressed
4 a.m.				dropped Heather at day-care
5 a.m.				work at bakery ↓
6 a.m.	woke up			
7 a.m.	exercise/coffee			
8 a.m.	work on computer ↓	slept late/day off		
9 a.m.		brush teeth		
10 a.m.		eat breakfast		
11 a.m.	interview for job	shower		picked up Heather
Noon		watched television		clean up and napped
1 p.m.		talked to friend	return home	↓
2 p.m.	talked to Lisa	talked to John	have lunch; play with Heather ↓	played with Heather ↓
3 p.m.	shower	studied Lamaze	and nap	
4 p.m.	study ↓	materials		
5 p.m.			cleaning/ vacuuming	made dinner
6 p.m.	television telecourse		watch television	showered with Heather
7 p.m.	relax with Lisa ↓	cooked dinner		watched television
8 p.m.			bed ↓	put Heather to bed
9 p.m.	visit sick father-in-law in hospital	visit sick father in hospital		watched television ↓
10 p.m.				
11 p.m.	home and bed	home and bed		bed

arose, and Lisa did not interfere or try to take over. He made some interesting points when asked if he felt Heather had limited his career opportunities:

> I still have plenty of time to do what I need . . . we already have responsibility and things we have to do, so you still find time to do what you have to do, it hasn't stopped me from my career opportunities, I'm still going to school . . . I think that other things might stop your career, a kid could as well. If you're that kind of person that, you know, it is a

> finality! In our society there are so many ways that, you know, help . . . It is not like the old days, you know, and [today] men and women [are] together in a relationship that is much more flexible . . . You have to be laid back too, I think you have to be kind of relaxed, not take everything so seriously.
>
> Thinking about becoming a parent? My advice would be, don't make having a child ... any more important than anything else you do in your life. It is not something to be afraid of, and it's, it's really easy after you do it: child-care and taking care of a child. It's really not hard. It's something that you do. People put too much influence on all that stuff, 'oh, it's very hard', and a lot of the times, I think people tried to scare us, I think. I mean, we feel fine, we don't have any real problems with her. The joy of having a kid is so much greater than any of the responsibilities. It is not an even exchange at all.

Lisa overheard this latter part of John's interview from the adjoining room and started to laugh. She confided to her interviewer, out of earshot of John, that she was shocked he felt their lives were so easy and had changed so little. She noted that, unlike John, she felt she was unable to return to school and that her career as a baker was limited because she now could only work part-time (see Table 7.1). Such radical changes, she noted in confidence to the interviewer, might suggest who in the family is most affected by child-rearing and who, in actuality, takes most responsibility for Heather and the household.

Nonetheless, for the most part, Lisa was satisfied with the balance between her work and home, but the negotiation of her mothering was tempered by the elder influences that John alluded to in the first set of interviews:

> I feel conflict between being a parent and working, because there's a lot of stigmatism given to a working mom, I think, especially from the older folk. They want you home all day with your child. But you just can't do what 'mom' says anymore. I feel that conflict some of the time. It's the other, you know, like family, saying that you should stay home or hire someone to come in and help you.

One of the things Lisa had hoped for was that the birth would bring her closer to John and that their dual responsibilities would establish new bonds of intimacy. In actuality, her experience, like many of the other parents we interviewed, was quite different. The irony is that the lack of time Lisa feels that she and John have for themselves as a couple is replaced by a perceived closeness with the rest of her family:

> There are less intimate times (laughs). There's less closeness to John, I mean as far as going out and being alone. She's at an in-between age where you can't take her on vacation with you, so there's less travel time. It's just different. I'm closer to my family, I think, because of her. I see them much more often. I'm tired more (laughs), but I'm more fulfilled, so it's like I'm whole. I don't know, it's different now. I don't know if it takes a child to do that, but it did!

When probed on this response, Lisa's discussion takes a different turn as she reflects on how her mothering transforms time and space:

> Oh, it's just more time consuming than I ever, ever imagined it would be. I thought it would be, you know, I don't know, we'd have strolls in the park and, you know, I'd bake cookies (laughs). No, (laughs) we buy them from the bakery.

One of the central assumptions of this chapter is that fatherhood and motherhood are morality plays that derive from cultural norms that constrain as much as they enable. As a consequence, parenthood has little to do with moralities embodied in the experience of parenting. The above interview excerpts suggest that, for some young parents, patriarchal gender roles and relations are engendered in fatherhood and motherhood to the extent that life with children – although immensely satisfying – is also enervating. The question remains, then, as to where precisely the child-rearing rite of passage leads? As Lisa notes, parenting is work and there is little theory to help us understand how space is produced and time contextualised by this kind of work. What encompasses the social imaginary of some young parents is a family fantasy that closely resembles the nuclear family with all its trappings of cookies, apple pie, mom in the kitchen and Mr Mom helping out when he can.

Parenting may be work, but we derive a large part of our political identity from parenthood and it is with this larger scale concept that patriarchy reifies itself through mythic structures and social norms. Like John and Lisa, Chad and Amy were in conflict over the work of parenting. Both made it clear that household duties after Amanda was born were shared equitably. Amy pointed out that Chad's help was important to her sanity and that he was, indeed, very helpful. Chad indeed worked hard around the house and with Amanda, but he made it clear that Amy orchestrated the domestic realm:

> Mm . . . there's a million unexpected things that may come up, like, if the baby's sick. Amy will call me up [at work] to do something for the home. Yesterday she probably called me a couple of times to make sure I get the bathtub crack sealed. Household responsibility!

The point is that Chad's domestic work is not necessarily about household responsibilities, but about 'helping out'. Like many couples we spoke to, Amy takes care of responsibilities whereas Chad helps out. For example, Chad may drop Amanda off at day-care but it is Amy who researches the long-term care needs of their daughter. Of course, it may be argued that Chad's and John's personalities and world-views are simply more carefree and relaxed than their wives. As Chad pointed out when asked what advice he would give new fathers: 'My father always said, the key to life is that there are two rules: one, don't sweat the small stuff; two, it's all small stuff. Try not to take everything too seriously, everything will work out at its own speed'.

I am not suggesting that Chad and John are shrugging their parenting responsibilities on to their partners, but their stories suggest certain feelings of patriarchal comfort. Chad, in particular, seems conscious that being male offers certain privileges. When, during our second set of interviews, one of our interviewers probed why nothing about Amanda seemed to surprise him and why he was so relaxed about his parenting he quickly responded to both questions with 'Well, because I am a man' and then laughed to signal he was joking. A mythic conceptualisation of fatherhood and complex emotions around fathering are exemplified by this quote from Chad when asked about whether having Amanda had limited his career:

> I think actually being male they [my employers] kind of expect me to have a child . . . eventually. Having a kid actually kind of opens a lot of doors for you because it makes you look (long pause) responsible and part of society rather than just some flake on the pavement. That may not necessarily be an accurate image but, ugh, it just kind of makes you look more down to earth as far as society is concerned. You look more established. It may also make you more of a captive audience. You're less likely to tell the boss get lost if you're afraid of losing your job. That too may be a false impression. It may be exactly the opposite. I may be more inclined to say: 'Forget it, I'm leaving to take care of my baby'.

No women in our study hinted that motherhood gave them such power over an employer. Like Lisa, Amy's feeling about parenting, for the most part, focused on the work involved. When asked what advice she would give someone who is thinking about becoming a parent, she retorted that you needed to make sure of an understanding employer and a co-operative partner. In addition, it was important that you not be afraid to ask for help. When asked how life with a new child was different from what she expected, she replied that is was much more time and energy consuming and then, almost as an afterthought, she noted that it was also more satisfying.

Conclusions: the rite of parenting

Rites guard the threshold between social categories of being and there is an assumption of fulfilment and a gained wholeness with a successful transition from one stage to another. In this sense, the mythic rites of motherhood and fatherhood foster political identity formation and social placement, but they do not reflect the day-to-day work of parenting, nor do they anticipate adequately the complex changes of daily living that accompany child-rearing. Although many of the parents in the San Diego study thought of motherhood and fatherhood as 'natural' categories that gave meaning to their emotions and behaviours, further scrutiny suggests that emotions and behaviours are complexly woven, unravelled and sometimes detached from the work of parenting.

Social theorist and pragmatist John Dewey (1957, 1989) noted that morality should be born from experience and that there is continuity and wholeness in day-to-day experience that promotes a structure from which morality should emerge. The linear sequence of rites of passage, and the propensity of transition-rite theorists to classify through large-scale dichotomies and hence to simplify the complexities that encompass human acts and responsibilities, are two issues that lie at the core of my concerns with contemporary wisdom on child-rearing rites. As William Golding points out in the opening epigraph, the scale is all wrong. My main point is that it seems that rites of passage into contemporary Western parenthood are simply rites to apply large-scale – 'gigantic', in Golding's words – patriarchally based covert cultural norms in the classification of motherhood on the one hand and of fatherhood on the other. The rite of *parenthood* is about a change that promotes inequitable gender relations and unworkable dualisms. But my research indicates that the practices – the rites – of *parenting* are very different: they are about a continuity that weaves work and practice with emotion.

By stressing both continuity and change, Dewey provided a radical attack on philosophical dualisms which were the basis of thought in his time and are still prevalent in social thought today (Cutchin 1999). Experience, writes Dewey (1989: 19), is a comprehensive activity and an inclusive integrity. Experience need not augment old patriarchal dichotomies when the work and emotions of parenting are joined. The experience of a first child will most likely result in some kind of transformation of gender roles and relations, and a re-negotiation of the sexual division of space and time. The spatial nature of these changes, and the fundamental questions of identity which arise from them, confront young parents to varying degrees such that they form a constellation of issues which define the diversity of family geographies. The mythic fulfilment of parenthood belies the day-to-day work of parenting and there is an inequitable gendered division of labour at the scale of the day-to-day, the local and the body that should (but rarely does) guide larger social and political imaginaries.

Acknowledgements

Research for this chapter was supported in part by Grant SES-9113062 from the National Science Foundation and a grant from San Diego State University Foundation. Special thanks go to the students who were employed on the study: Marta Miranda, Chris Carter, William Granger, Leslie Bolick, Suzanne Michel, Thomas Herman, Katina Pappas, Nickolas Deluca, Meg Streiff, Susan Mains, Matt Carroll, Serena McCart and Pauline Longmire. In addition, I would like to thank all the participating families. Any opinions, findings and conclusions or recommendations expressed in this chapter are mine and do not necessarily reflect the views of the National Science Foundation, San Diego State University or the students and families involved in this project.

Notes

1 The initial participant sample, when compared with San Diego census data, had a slightly lower than average income and slightly higher incidence of single-parent households. All participants were interviewed separately at a time and place (usually the home) of their convenience.

2 That is, conversion into everyday practices and structures of power.

3 Arguments for understanding scale as a social construction are sufficiently rehearsed in recent geographic literature to the extent that scale itself is no longer considered a 'natural' or unproblematic category (cf. Smith 1992, 1993; Herod 1997; Lukinbeal and Aitken 1998). In the same sense that fatherhood and motherhood are social constructions, then so too are issues that relate to space and scale (Aitken 1998).

4 Fernand Lamaze was a French obstetrician who developed a method of childbirth in the 1950s that involves psychological and physical preparation by the mother in order to suppress pain and facilitate delivery without drugs. Lamaze classes are the most popular form of childbirth preparation in the USA and often involve the mother and a partner.

References

Aitken, Stuart C. (1994) *Putting Children in their Place*, Association of American Geographers: Washington DC.

—— (1998) *Family Fantasies and Community Space*, New Brunswick, New Jersey: Rutgers University Press.

Barthes, Roland (1972) *Mythologies*, New York: Hill and Wang.

Campbell, Joseph (1949) *The Hero with a Thousand Faces*, Princeton, New Jersey: Princeton University Press.

—— (1972) *Myths to Live By*, New York: The Viking Press.

Cutchin, Malcolm (1999) 'Qualitative explorations in health geography: pragmatism and related concepts as guides', *The Professional Geographer* 51, 2.

Del Castillo, Adelaida (1993) 'Covert cultural norms and sex/gender meaning: a Mexico City case', *Urban Anthropology* 22, 3–4: 237–258.

Dewey, John (1957) *Human Nature and Conduct: An Introduction to Social Psychology*, New York: The Modern Library.

—— (1989) *Experience and Nature*, 2nd edn, LaSalle, Illinois: Open Court.

Golding, William (1980) *Rites of Passage*, New York: Farrar, Straus and Giroux.

Grimes, Ronald (1995) *Marrying & Burying: Rites of Passage in a Man's Life*, Boulder, Colorado: Westview Press.

Herod, Andrew (1997) 'Labor's spatial praxis and the geography of contract bargaining in the US east coast longshore industry, 1953-1989', *Political Geography* 16, 2: 154–169.

Homans, Hilary (1994) 'Pregnancy and birth as rites of passage for two groups of women in Britain', Carol MacCormack (ed.) in *Ethnography of Fertility and Birth*, 2nd edn, Prospect Heights, Illinois: Waveland Press, 221–258.

Kahn, Robbie (1995) *Bearing Meaning: The Language of Birth*, Chicago: University of Illinois Press.

Kitzinger, Sheila (1994) 'The social context of birth: some comparisons between childbirth in Jamaica and Britain', Carol MacCormack (ed.) in *Ethnography of Fertility and Birth*, 2nd edn, Prospect Heights, Illinois: Waveland Press, 171–194.

Lukinbeal, Chris and Aitken, Stuart C. (1998) 'Sex, violence and the weather: male hysteria, scale and the fractal geographies of patriachy', in Steve Pile and Heidi Nast (eds) *Places Through the Body*, London: Routledge, 356–380.

MacCormack, Carol (ed.) (1994) *Ethnography of Fertility and Birth*, 2nd edn, Prospect Heights, Illinois: Waveland Press.

Mead, Margaret (1955) *Male and Female: A Study of Sexes in a Changing World*, New York: New American Library.

Meade, Michael (1996) 'Rites of passage at the end of the millenium', Louise Carus Mahdi, Nancy Geyer Christopher and Michael Meade, (eds) in *Crossroads: The Quest for Contemporary Rites of Passage*, La Salle Illinois: Open Court, 27–33.

Paige, Karen Ericksen and Paige, Jeffrey M. (1981) *The Politics of Reproductive Ritual*, Berkeley: University of California Press.

Ruddick, Sara (1992) 'Thinking about Father', in Barrie Thorne and Marilyn Yalom (eds) *Rethinking the Family: Some Feminist Questions*, Boston: Northeastern University Press, 140–154.

Smith, Neil (1992) 'Geography, difference and the politics of scale', in Joe Doherty, Elspeth Graham and Mo Malek (eds) *Postmodernism and the Social Sciences,* London: Macmillan, 57–79.

—— (1993) 'Homeless/global: scaling places', in Jon Bird, Bary Curtis, Tim Putnam, George Robertson and Lisa Tickner (eds) *Mapping the Futures: Local Cultures, Global Change*, London: Routledge, 87–119.

Van Gennep, A. (1960) *The Rites of Passage*, trans. M.B. Vizedom and G.L. Caffee, London: Routledge and Kegan Paul. First published in 1909, *Les Rites de Passage*, Paris: Noury.

Young, Frank W. (1965) *Initiation Ceremonies: A Cross-cultural Study of Status Dramatization*, New York: Bobbs-Merrill.

8

WOMEN'S EXPERIENCES OF VIOLENCE OVER THE LIFE-COURSE

Rachel Pain (UK)

Experiences of violence and abuse do not constitute 'rites of passage' in the sense of relatively predictable and patterned life events commonly experienced at particular life stages. Yet, in common with the major life events described elsewhere in this collection, violence often brings significant changes and adjustments to individuals' sense of their identity and the geographical world around them. This chapter focuses on women, for whom experience or awareness of the threat of aggression contributes to the learning processes involved in growing and ageing, and to a greater or lesser extent affects the way we live, perceive and manage our selves, our bodies and the spaces we use.

In particular, the threat of invasion or damage to the body through sexual violence leads to restrictions on many women's use and perception of different spaces (Hanmer and Saunders 1984; Valentine 1989). The experience of rape itself, perhaps the ultimate domination of another's body, leaves emotional scars which take years to heal. The incidents related by women in this chapter include the extreme violent events of rape, child sexual abuse and domestic violence, but also more commonplace experiences of harassment, defined here as unwanted and intrusive behaviours which cause a feeling of threat or humiliation, such as verbal abuse, flashing and stalking. Harassment may also be seen as part of the embodiment and subordination of women (MacKinnon 1979; McDowell 1995), usually involving verbal or physical reference to bodily difference, and an implied threat to the integrity of the body. Responsibility and blame for both sexual violence and harassment are linked to broader cultural codes about women's bodies (Brownmiller 1973; Russell 1982).

Violence has been widely viewed in the feminist literature as an effective device for regulating the female body, and the process of keeping physically safe is one mode of 'performative femininity' (Stanko 1997). Experiences of violence and threat, coupled with knowledge of the experiences of others, forge senses of vulnerability and confidence which change over the life-course, are situated geographically and culturally, and are embodied: 'I wouldn't have walked home through that park at night, looking like that'; 'I would have thrown him out before he did that to me'. Equally, experiences change our perceptions of different space as safe or dangerous, and our strategies in negotiating them (Stanko 1987; Painter 1992).

Recent work has emphasised women's active resistance, both to violence and to the social expectation of fearfulness (Koskela 1997; Stanko 1997). Survivors of violence react to their experiences, grow and move on. I explore some of these issues in this chapter, focusing on individual stories from women at different points in the life-course in order to situate violent incidents, harassment and fear in the broader context of life changes. Through a discussion of some of the impacts violence may have through childhood, adolescence, adulthood, motherhood and old age, my aim is to illustrate the connections and repercussions between these stages of the life-course, the cumulative development of skills and attitudes to danger which women acquire, and the ongoing processes of growth and change in identity to which violence and its threat contribute.

The Edinburgh study

The material used here is drawn from research carried out in Edinburgh, Scotland in 1992, which investigated a number of aspects of women's experiences of violent crime. A mail questionnaire survey was carried out with 389 women, the results of which are summarised elsewhere (Pain 1997). Material drawn from follow-up in-depth interviews forms the basis of this chapter. Interviews were carried out with forty-five women in their own homes, and lasted between one and three hours. They were tape-recorded with permission and later transcribed. Interviewees were recruited from three study areas within 5 kilometres of the city centre, which provided a sample of women from a broad range of social and economic backgrounds. Amongst the women, then, are very different residential contexts, and also wide variation in age, social class, ability, marital status and so on. The women were all white, and only two interviewees identified themselves as lesbian. While these different social identities influence the nature and meaning of experiences of violence (Pain 1997), the continuities in experience across these boundaries is also striking. The personal stories related here are not presented as typical, but are chosen to illustrate the commonness of violence, harassment and concern about them, and the ways in which they shape women's lives.

Children: the quintessential victims

Walklate (1989: 68) describes children as 'the quintessential victims: they are structurally powerless as well as physically powerless'. As we are frequently reminded, children may also be the perpetrators of crime, but they are equally or more likely to be victims (Hartless *et al.* 1995). At this time, girls are more vulnerable to sexual and physical violence than at any other stage in the life-course. A quarter of the women who responded to the mail survey reported experiencing an incident of sexual or physical violence (as distinct from harassment) before the age of 16. These incidents, and the warnings about danger which parents give to young girls, provide influential early learning experiences. Experienced at any point in the life-course, violence changes lives. In childhood it may have particularly traumatic and long-lasting repercussions.

Barbara is an interviewee in her thirties who at the time of the research was separated from her husband, with two children. She has lived in fear for over twenty years after going to court to help convict the man who raped her when she was 12, and describes the effects that this has had.

> When I was found out and I was taken to a police station, and in a cell to get examined by the police doctor, that was like another invasion of my body. And um I had to go up on the stand and literally relive it, oh it's a nightmare – Now when it happened to me my hair was down there [waist-length] and I still had long hair when my mother showed me the piece in the paper that he got let out. I had all my hair chopped off, I was terrified he would come and look for me – He is 60 years old now. He was 32 when it happened to me. And he's still doing it.
>
> (Barbara)

The sense of lack of support from the criminal justice system is not unusual amongst survivors of abuse. Barbara's feelings of guilt about what happened are also common; as a 12-year-old she blamed herself for the rape, changing her appearance when the rapist was released from jail three years later, in an attempt to make herself less physically attractive. The emotional harm continues to affect every aspect of her personal life as an adult.

> It can be damaging in a lot more ways than what people think. It's like an invasion of your body obviously, and you just cannae trust anybody. And obviously I've never been able to trust anybody long enough. And I joke about it, och aye I've been three times married, I just like the wedding cakes, you know. But that's just me. Because I have to make a

> joke out of it. Because I don't want everyone to know how I feel deep down inside.
>
> (Barbara)

Common long-term effects of child physical and sexual abuse, including depression, lack of self-confidence and difficulty establishing trusting relationships, are well documented (Parker and Parker 1991). However, at the same time survivors grow and move on. In this study, the women who had been sexually abused as children felt less fearful about violence as adults than those who did not report abuse. They feel what happened to them as children was totally beyond their power, but they are more confident about exercising choice and control about adult relationships. The following written comments from three different women illustrate this point: 'Have more confidence and able to face up to people', 'Don't see the same situation arising as when I was a child', 'I have left home and am now married'.

For some, it is when they have their own children that these incidents become significant once again. This may mean the return of vivid, unwelcome memories; one woman received counselling for the first time after depression brought on by the birth of her son. Amanda, another woman in her thirties, was abused by her uncle when she was a child, and now has four children of her own aged between 2 and 16 years. She goes to great lengths to ensure that the same does not happen to them, at great personal cost.

> I'm paranoid about my children. Even with workmen in the house, and if we go to visit anybody I'm on edge – It's just you're no, you're no confident now, I mean it's no just physical underneath, it's what you go through mentally – Certain people that come in, I mean my older laddie's friends come in and I think 'no I dinnae like him', I mean the hair on the back of your neck stands up and that makes you all agitated. And it's stupid. It's stupid. But it's there all the time. I cannae get away from it.
>
> (Amanda)

Adult fears are the result not only of first-hand experience, but of a long process of learning and socialisation which begins in childhood (Goodey 1994). While most of us develop skills to cope with physical and sexual danger over the life-course, children are frequently not in a position to prevent abuse, or even report abuse to other adults. This is not because parents today and in the past did not warn children about danger; in fact, accounts from women of different ages in the study shows this to have been an historically consistent element of lessons given to girls about the environments around them. However, warnings were and are often quite vague and unspecific, with strong spatial messages centring on public space. These warnings rarely match children's experiences of abuse and harassment in private space.

OLIVIA: It would be like, don't talk to strangers, don't ever get into a car with strangers, don't accept sweets.

RP: Did they ever say why?

OLIVIA: Um, probably that they would maybe kidnap you or something, you know they would take you away. But there would never be anything more, um, not sordid I suppose, nothing deeper explained.

> I was warned as a youngster never to go into anyone's car. But never about being attacked or anything like that. No she just says just don't ever go in a stranger's car. But never ever told you what would be done to you, you know.
>
> (Gail)

> I was sort of fondled as an adolescent by the lollipop man who knew my mother, spoke to my mother, was allowed into the school. You know, trusted position. And I didn't report that because I was in such a dilemma because he wasn't a stranger. It threw everything to one side.
>
> (Sheila)

> Don't go with strangers is all I remember – It meant that when you did have sort of groping uncles or friends, you felt uncomfortable but nobody had said, you know, this is not on, you should do something about it. Because they'd never been specific.
>
> (Elaine)

Parents' fears and warnings about stranger danger have a restrictive effect on children's geographies (Pain 1994), justified by the cultural image of children as 'vulnerable and incompetent in public space' (Valentine 1997). For girls, these warnings and restrictions teach important messages about the body and responsibility for its safekeeping (Stanko 1990), especially as they become teenagers.

Adolescence: learning heterosexual identity

Most women learn a clear sense of their sexual vulnerability in adolescence (Burt and Estep 1981) alongside the physical changes experienced at this time. Warnings from parents increase and become more clearly focused upon sexual danger, and experiences in public places and the home may reinforce these as girls grow older. Adolescent girls are particularly vulnerable to sexual harassment, which has a role in the construction of gender and sexual identities, in public and private places (Lees 1986; Wise and Stanley 1987; Herbert 1989). Almost two-thirds of the women in the mail survey report an incident of sexual harassment having occurred before they were 16.

> A friend and I used to go together to a club, at that time I stayed near Dalry Road, and we went down what they call Coffin Lane. And a man followed us down there. And we ran. And luckily when we got to the bottom, we got to Dalry Road, and he went away. But um we really got a fright then. We used to go down there every time, you know, and it never bothered you, but once that happened well you didn't because it frightened you.
>
> (Maureen)

> When I was a teenager a boy who was a bit older than me got me in a corner and started touching places he shouldn't have been touching. And that again I thought was my fault – I mean it was a church, a youth fellowship sort of thing, you know, I mean you just don't expect things like that to happen there. And I thought, well that must have been my fault for getting into that position.
>
> (Valerie)

Learning to deal with harassment in a way that women feel is effective and assertive can take years longer. For many, self-imposed spatial constraints, social guardedness and careful regulation of bodily appearance become part of female identity (Gardner 1990), and girls growing up learn that:

> The Good Woman – and this suggests the law-abiding, middle-class, modest, risk-averse – will not walk down dimly lit alleyways, forget to put petrol in her car, carry her handbag close, hitchhike and will hide her jewellery when in public – Women who do not follow the rules for prudent behaviour, it is presumed, deserve to be excluded from any benefits of public provision of safety, because those women (not us) fail to take appropriate measures to protect themselves from harm.
>
> (Stanko 1997: 486)

Women are thus encouraged to adopt 'feminine' behaviour for their own protection by parents, teachers and official crime prevention agencies, and to take care to avoid behaviour or an image which might appear sexualised in inappropriate settings. By late adolescence, this caution may be self-directed.

> My attitude towards men has changed as I've got older. If say I'm going out with someone who I may not know very well, or even if I know him, then I am careful, in making sure we go to a crowded place and I'm not left on my own with him – I think you have to really, because at the end of the day, and all my friends will agree on this, at the end of the day it's up to you. That's the impression that you get. It's up to you to be careful.
>
> (Marie)

> I do like to dress up and wear a short skirt and high heels but I won't do it unless I'm in a big group – It's not so much that I worry it would cause someone to rape me but if it did happen, then it would be felt by the judge and the police and whoever that I had caused the attack in some way.
>
> (Elizabeth)

Heterosexuality contains danger but also, ostensibly, safety, if proper behaviour in the right places is observed, based on the notion that acquiring one particular man will protect women from all other men (Valentine 1992). For women who do not fall into this category, their sexuality also carries danger and safety. Caroline and Janet are a lesbian couple living together, both now in their thirties. Although they generally feel safe from violence today, they describe being victimised as teenagers, a common experience amongst girls and boys who appear to challenge heterosexual norms (Lees 1986; Harry 1992).

CAROLINE: Getting called names, being physically mauled, sort of physically attacked, I remember them always pulling my hair and trying to pull my earrings out, that sort of thing ... I got sort of constant verbal abuse off some of them, all the way home after school every day. That went on for years. I just walked along and ignored it.

RP: What sort of names did they call you?

CAROLINE: Oh – lezzie, dyke. Whore. Which is a bit mixed up! [laughs]

RP: Did you experience similar things?

JANET: Yes. Yes I did. In fact I remember one teacher sort of taking me to one side and saying, you know, really right on, well who you are is up to you, but wouldn't it be better if just now you didn't dress so outrageously, and just like I should try and pretend to be normal . . . Being gay is presented as this great danger, but you know you can see around you that the biggest danger is not being gay, if you're a woman. Generally speaking you're safer than anyone else, in that sense.

CAROLINE: Well maybe not as a teenager, but definitely when you get older.

Adulthood: turning points and resistance

For many women, early adulthood, particularly leaving home, provides some freedom from the constraints which parents may expect. It may also be a time of increasing awareness of danger which comes for some with new experiences of violence and harassment in public places, the work place or home. Of these locations, the home is the most common site of violence (Johnson and Sacco 1995; Mirrlees-Black *et al.* 1996), but despite this, knowledge about domestic violence in Western societies is paradoxical. On the one hand public awareness has improved dramatically, and on the other it tends to be distanced by individuals from their own situation (Pain 1997), partly due to the emphasis of public discourses on stranger danger. Reactions to domestic violence also stem from deeply entrenched cultural notions attributed to those who suffer it: they may be viewed by others as somehow to blame, unable to treat men properly or deal with them assertively (Duncan 1996). All this means that the experience of violence from partners can come as a great shock, rocking personal feelings of physical security as well as aspirations about marriage and partnership.

RP: Is [violence in marriage] something you expected when you were single?

VALERIE: No, you expect it to be love and er, no, everybody's brought up with this idealised image that it's all going to be love.

RP: Did you ever imagine anything like that could happen when you were younger?

DIANA: No. Nope. Never.

RP: Did you know about violence in marriage?

DIANA: I never knew about violence in marriage until, like, my friends started to get married. 'He done that to you? I'd kill him if he done that to me!' You know, and I was pretty old when I got married, I was 28, but um, no I never thought he could be like that to me.

Diana, now in her forties, experienced violent behaviour from her husband as soon as they were married. It culminated a year later, when she was six weeks pregnant – her husband woke her up in the middle of the night and subjected her to a violent sexual attack at knife-point. Diana managed to escape to her parents' house, where she suffered a breakdown and was unable to speak for three days. Like those who suffer abuse in childhood, the violence has had a lasting and damaging effect. Ten years on, Diana's story is of overcoming her fear and gradually acquiring the strength to stand up to her ex-husband's continued intimidation.

> He actually walked in four months ago. I was very cool calm and collected about it – he taught me patience. Because when he did that to me, I thought you are never gonna get near this child of yours. You are bad. And I'm gonna have to wait my turn and fight this out, and I'm gonna win – He said you're not the person I knew, and I said I'll bet, I'll bet. – Why are you so cool? – I says I'm not afraid of you any more.
>
> (Diana)

One effect of suffering violence in private space is that, at least in the medium term, wider feelings of security may be shattered. The women in the research who had been beaten or raped by husbands or boyfriends in recent years report that these experiences have heightened fear of sexual attack and their use of precautionary behaviour in public places, as well as making it more difficult to trust men they know.

> Well I'd keep to a place that's a very, er, lit up area. I'd never think about taking a short cut down an alleyway or anything like that. Never. Years ago I would have, but not now – I'll tell you something, it made me dead wary of men. I can read, I'm quite good at reading people. I dunno if it's sorta wisdom, you know, or experience or whatever.
>
> (Diana)

The risks of attack in public space are at least perceived as reducible, either by staying inside or by using a range of coping strategies to negotiate them. The threat of violence from relatives or partners is not so easily bypassed, though many women do employ numerous behavioural strategies in an attempt to reduce the chances of danger and to resist it when it occurs, including verbal negotiation and the avoidance of 'trigger' situations (Dobash and Dobash 1992). Valerie is in her thirties, married with two children, and describes herself as a victim of marital rape. Over the years, she has developed various ways of trying to cope, both psychologically and tactically.

RP: Does it affect your behaviour towards him?

VALERIE: Oh yes. I would say so. I will do anything to avoid a scene where he demands sex. Just as he uses any excuse to sort of cajole me into it – It's not sort of something that can be avoided when it's coming from someone you live with. Put off yes, perhaps – And especially having

> children, I think that if you're just on your own and you don't have any commitments, for one thing you can just walk out the door but also you can sort of shout and yell and scream and stand up for yourself.

The term 'fear of crime' has been associated particularly with negative and passive reactions to the threat of (largely public space) danger, but women's emotional and practical reactions to the possibility of violence are wide ranging. Most women exhibit spatial confidence as well as spatial restriction and fear (Koskela 1997). These apparently diametrical senses of confidence and vulnerability are not mutually exclusive, but are features of all our lives to different extents, influenced by spatial, temporal and social contexts. Women do not simply accept the assignation of actual or potential victim, nor get progressively more fearful as they age, but emotions and responses alter according to events. For Koskela (1997), women's life-courses are characterised by these 'breakings' or turning points. One feature of adulthood for most of the women in this study is a growing sense of resistance to fear of stranger danger.

> I have avoided going out because I was too scared but usually if I'm frightened of something I force myself to do it. To overcome the fear basically.
>
> (Karen)

> I mean I try not to curb my lifestyle at all – It's kinda like, you know, refusing to cross the road in case you get knocked over.
>
> (Jane)

> It doesn't have any effect on me. Nah, it doesn't. Or rather I won't let it affect me. I'm no staying in because of one or two maniacs running around out there, I'm not gonna change what I do, I go out at night, I walk down Ferry Avenue, I do as I please – Your chances of anything happening are so wee so you've got to put that into perspective.
>
> (Jackie)

Equally, and again in contrast to the popular image, women show resistance rather than passivity when dealing with harassment. In adulthood, many feel they can deal with flashing, sexual comments and so on received from men in public places, and are less worried by it as they get older. A new location of harassment, however, is the work place. Half of the women in the mail survey reported unwanted, intrusive sexual behaviour from men in a shared work place, and this comes as a considerable shock when it is first experienced.

> Basically there was a huge stationery cupboard and I wouldn't go in it with him [laughs]. And he'd make sort of comments and I didn't care

> for it. Um but I left the job. If it was the same job with a different boss I would have stayed in it.
>
> (Vicky)

KAREN: When I first left school I worked for Edinburgh Council er and I was sexually harassed there, and because I slapped a person in the face that had sexually harassed me I was sacked.

RP: What sort of things was he doing?

KAREN: Nipping my bum and things like that. And because I slapped him in the mouth I was fired.

Work place harassment has been described as an expression of the male-dominated and heterosexual nature of many work spaces (MacKinnon 1979), one of the ways in which female bodies 'out of place' are disciplined (McDowell 1995). Its impact varies widely, according to the job, the working context and the resources available to contest it. However, all of the women interviewed who had experienced harassment had developed coping strategies.

RP: Does it ever get to the point where it's worrying you?

BARBARA: No I just hit them. I give a couple of guys a belt for giving me some harassment. And as I said, having a big family, there's that many of us and I just let it be known about the big family. If things got out of hand I'd just have to speak to them and they'd sort it out for me.

> I think sometimes if you turn it back on them, if you make them feel uncomfortable, telling dirty jokes and whatever. The janitors and that, they try and be really smart with you, and if you turn it back on them they get embarrassed. Cos they dinnae ken how to cope with it.
>
> (Amanda)

Several say that as they got older, they learnt how to pre-empt these events, though like domestic violence or abuse from strangers in public parks and streets, the common implication is that women are responsible for preventing and dealing with harassment. While there are many stories of successful resistance, and some incidents of harassment can be a cause of little concern and even hilarity, the view that resistance only constitutes 'fighting back' reflects a dichotomy of oppression and liberation which serves the interests of the powerful (Wise and Stanley 1987), and is unhelpful to those who suffer the most.

Alex, in her twenties, and Elaine, in her fifties, both work in male-dominated professions. Both feel they experience hostility from the men they work with because of their own relatively senior positions. This hostility is regularly expressed in rude and malicious remarks about Alex's appearance, sexuality and ability to do her job, and a lack of co-operation on work tasks, while Elaine feels that her ideas and views are often overlooked or trivialised because she is female. She has also had problems with other senior members of staff making sexual overtures towards her and who, once rebuffed, bore a grudge and made it more difficult for her to do her job. In both cases, harassment has had a serious effect on their enjoyment of work and confidence in their own abilities, and at the time of the interview Alex was considering changing career while Elaine looked forward to retirement.

> People whistling at you and things like that, I could take in a light-hearted way. But there's a maliciousness involved at work in comments about my appearance and my abilities, a distinct personal maliciousness that is definitely well out of the bounds of anything humorous or whatever – I thought that I was strong enough to cope with staying in the industry and that I would be able to personally, you know, make them change things, and I was strong enough to survive, and that whatever anybody else said I would keep going. But it really does get you down after six years of it. And I'm beginning to wonder is it really still worth fighting so hard just for the right to do your job as well as they do?
>
> (Alex)

> Well it's a terrible combination of anger and misery really. I mean you know this [one incident] happened about a year ago, and as I say I will still have no dealings with this guy, I won't speak to him or even look at him, and I think that's a ridiculous situation. [Harassment] is being used as a way to avoid equality. I mean it's one thing that they realise they can make women upset. It's one way of getting ahead.
>
> (Elaine)

Old age: strength within

Older people in general, and older women in particular, are often represented as unduly fearful about violent crime (Midwinter 1990). In the study reported here, not only do levels of fear amongst women show no increase with age, but it is clear that cumulative lived experience can alter older women's attitudes to safety in more positive ways (Pain 1995). Women over the age of 60 are not a separate entity from other adults, and nor can their present-day emotions and reactions be disaggregated from events experienced and knowledge gained over the rest of their own life-courses. In addition, although the bodily changes which accompany old age may lead to some feeling more frail and vulnerable, the notion of physical vulnerability cannot exist in a vacuum, but necessarily has social meaning – bodily changes make you vulnerable to what? – from whom? – and why? While the socially constructed dependency resulting from physiological change in old age in Western societies can explain one particular age-specific form of violence, elder abuse (Aitken and Griffin 1996), general risks of violence at this life-stage are no greater, and as a growing body of evidence suggests, neither is fearfulness (Midwinter 1990).

The women over 60 expressed less fear than younger women, and they are also less likely to curtail their day-to-day activities and mobility in response.

> I think if I had to stay in because of something like that [fear of attack] I'd be more determined to go out, you know. I can appreciate it with people if they're feart, but I think it makes me more determined to go.
>
> (Moira)

> That's part of the reason why my handbag is so heavy. A clonk from that would knock him out for – you know. [laughs] I always put an umbrella in the bottom of ma bag to make it extra heavy. Just in case I need it.
>
> (Sharon)

Such resistance might be explained by the possibility, that Edinburgh was safer when they were young. Older women living in the outlying areas of Pilton and Corstorphine recall walking home alone from the city centre in the early hours of the morning after a night out, behaviour which would be unthinkable to most young women today. Having seen the city become a more dangerous place relatively quickly, the sense of indignation amongst older women is greater. Young women who have grown up with a virtual curfew on using public space at certain times may question it less. Often the main fear older women expressed was for the safety of their daughters, sons or grandchildren rather than for themselves.

However, given that the older women were as likely as women in other age groups to report having experienced violence and harassment earlier in the life-course (including a slightly higher rate of domestic violence), the notion that women grow and move on from these experiences seems most evident here. In interviews, these women describe their age, knowledge and experience as contributing a positive mindset towards danger. Edna, in her seventies and now living in sheltered housing, feels strongly that rape and abuse were fairly common in the first half of the twentieth century, but were not talked about. Through finding the strength to deal with a dangerous situation herself, Edna acquired a greater sense of security which has remained with her throughout her life.

EDNA: I had a nasty experience when I was young, I mean it's not new to this age – I got into the taxi and then suddenly he turned off and he went along a narrow unmade road and we ended up on this bit of wasteland. I said are you going to attack me? He didn't answer you see. And I said well I'll scream at the top of my voice, that's the first thing I'll do if you touch me. And er he stood there and I said are you a married man? He said yes. And have you got children? He said yes. I said do you know what's going to happen if you do anything to me? I says you're going to lose your job, you're going to lose your wife, you're going to lose your children. And er he says I would pick you wouldn't I. He says jump in and I'll take you back. – You see then people didn't speak about things like that. I mean everything's more open now. But I mean it was something that we'd feel ashamed of then and you wouldn't speak about it.

RP: After that, did it make you feel more wary?

EDNA: No. No, no. I've always felt that I could hold my own anyway. It made me feel more secure, that I could deal with a situation like that.

Conclusion

Such recognition that experiences of danger are relatively commonplace has been accompanied by changes in the way in which survivors of violence are viewed by society; what was a private problem until relatively recently is now becoming a public admission in many countries. Testament to this is the openness with which a wide range of experiences were reported to this research. Several women had only recently told friends or relatives of incidents of abuse which had happened to them years ago, and others had been encouraged to seek out counselling in what they perceive as a climate of greater acceptance. Yet violence remains unpredictable, and shatters lives when it happens. It generates, both for survivors and those who have not encountered it, a sense of sexual and physical vulnerability that women react to in different ways as they get older.

Increasingly, the stories of men as survivors of sexual and physical violence, and the broader impacts of fear on men's lives and their experiences of masculinity, are also being told (Stanko and Hobdell 1993; Goodey 1997). What is rarely dwelt upon is the offender. For every survivor, there is a perpetrator who has knowingly or unknowingly wrought these transitions, turning points and adjustments in the lives of others. More positive societal attitudes to survivors do not reflect a change in attitudes amongst those who are violent, nor imply that they have been punished (very few of the incidents reported to this research had been reported to the police). Instead, the men and women who are threatened by their violence are ultimately those who deal with it, through processes of negotiation and change.

References

Aitken, L. and Griffin, G. (1996) *Gender Issues in Elder Abuse*, London: Sage.

Brownmiller, S. (1975) *Against our Will: Men, Women and Rape*, London: Martin, Secker and Warburg.

Burt, M.R. and Estep, R.E. (1981) 'Apprehension and fear: learning a sense of sexual vulnerability', *Sex Roles* 7, 5: 511–522.

Dobash, R.E. and Dobash, R.P. (1992) *Women, Violence and Social Change*, London: Routledge.

Duncan, N. (1996) 'Renegotiating gender and sexuality in public and private spaces', in N. Duncan (ed.) *Bodyspace: Destabilizing Geographies of Gender and Sexuality*, London: Routledge, 127–145.

Gardner, C.B. (1990) 'Safe conduct: women, crime and self in public places', *Social Problems* 37, 3: 311–328.

Goodey, J. (1994) 'Fear of crime: what can children tell us?', *International Review of Victimology* 3: 195–210.

—— (1997) 'Boys don't cry: masculinities, fear of crime and fearlessness', *British Journal of Criminology* 37, 3: 401–418.

Hanmer, J. and Saunders, S. (1984) *Well-founded Fear: A Community Study of Violence to Women*, London: Hutchinson.

Harry, J. (1992) 'Conceptualising anti-gay violence', in G.M. Herek and T.K. Berrill (eds) *Hate Crimes: Confronting Violence against Lesbians and Gay Men*, London: Sage, 113–122.

Hartless, J.M., Ditton, J., Nair, G. and Phillips, P. (1995) 'More sinned against than sinning: a study of young teenagers' experiences of crime', *British Journal of Criminology* 35, 1: 114–133.

Herbert, C.M.H. (1989) *Talking of Silence: the Sexual Harassment of Schoolgirls*, London: The Falmer Press.

Johnson, H. and Sacco, V. (1995) 'Researching violence against women: statistics from Canada's National Survey', *Canadian Journal of Criminology* 37: 281–304.

Koskela, H. (1997) 'Bold walk and breakings: women's spatial confidence versus fear of violence', *Gender, Place and Culture* 4, 3: 301–319.

Lees, S. (1986) *Losing Out: Sexuality and Adolescent Girls*, London: Hutchinson.

McDowell, L. (1995) 'Body work: heterosexual gender performances in city workplaces', in D. Bell and G. Valentine (eds) *Mapping Desire*, London: Routledge, 75–95.

MacKinnon, C. (1979) *The Sexual Harassment of Working Women*, New Haven, Connecticut: Yale University Press.

Midwinter, E. (1990) *The Old Order: Crime and Older People*, London: Centre for Policy on Ageing.

Mirrlees-Black, C., Mayhew, P. and Percy, A. (1996) *Home Office Statistical Bulletin 19/96: The 1996 British Crime Survey*, London: Government Statistical Service.

Pain, R. (1994) *Kid Gloves: Children's Geographies and the Impact of Violent Crime*, Departmental Occasional Paper, New Series No. 1, Division of Geography and Environmental Management, University of Northumbria.

—— (1995) 'Elderly women and fear of violent crime: the least likely victims?', *British Journal of Criminology* 35, 4: 584–598.

—— (1997) 'Social geographies of women's fear of crime', *Transactions of the Institute of British Geographers* 22, 2: 584–598.

Painter, K. (1992) 'Different worlds: the spatial, temporal and social dimensions of female victimisation', in D.J. Evans, N.R. Fyfe and D.T. Herbert (eds) *Crime, Policing and Place*, London: Routledge, 164–195.

Parker, S. and Parker, H. (1991) 'Female victims of child sexual abuse: adult adjustment', *Journal of Family Violence* 6, 2: 183–197.

Russell, D. (1982) *Rape in Marriage*, New York: Macmillan.

Stanko, E.A. (1987) 'Typical violence, normal precaution: men, women and interpersonal violence in England, Wales, Scotland and the USA', in J. Hanmer and M. Maynard (eds) *Women, Violence and Social Control*, London: Macmillan, 122–134.

—— (1990) *Everyday Violence: Women's and Men's Experience of Personal Danger*, London: Pandora.

—— (1997) 'Safety talk: conceptualizing women's risk assessment as a "technology of the soul"', *Theoretical Criminology* 1, 4: 479–499.

Stanko, E.A. and Hobdell, K. (1993) 'Assault on men: masculinity and male victimisation', *British Journal of Criminology* 33, 3: 400–415.

Valentine, G. (1989) 'The geography of women's fear', *Area* 21, 4: 385–390.

—— (1992) 'Images of danger: women's sources of information about the spatial distribution of male violence', *Area* 24, 1: 22–29.

—— (1997) '"My son's a bit dizzy." "My wife's a bit soft": gender, children and cultures of parenting', *Gender, Place and Culture* 4, 1: 37–62.

Walklate, S. (1989) *Victimology*, London: Unwin Hyman.

Wise, S. and Stanley, L. (1987) *Georgie Porgie: Sexual Harassment in Everyday Life*, London: Pandora Press.

9

LIFE AT THE MARGINS

Disabled women's explorations of ableist spaces

Vera Chouinard (Canada)

For women with disabilities, negotiating spaces of everyday life, such as the home and work place, is often a difficult, contradictory and oppressive experience. This is because experiencing spaces through a disabled body not only involves significant physical and mental challenges, dealing with significant limits to one's capacities to act, but also encountering and responding to complex, often confusing social rules and cultural codes which mark the disabled body as negatively different and less valuable than the 'taken-for-granted' (norm) of the able body. 'Cultural oppression' is the term used by Iris Young (1990) to refer to this social construction of difference as negatively 'other'. It occurs, she argues, through an array of discursive and practical reactions to those who differ from culturally dominant groups. At one extreme are conscious, overt acts, such as racist remarks. At the other are mundane responses to someone who is perceived as different or 'other' – responses such as ignoring their presence or reacting to them in ways that help to mark them as negatively different (e.g. singling a visibly disabled person out for unwanted attention).

In this paper, I explore some of the ways in which disabled women's places in society are shaped by processes of cultural oppression and how they contest those places through 'taking up space' in ways that challenge their oppression. I begin by sharing some personal journeys through ableist spaces: first, my own as a disabled woman in academia, and, second, those of disabled women activists in Canada. To negotiate ableist spaces of life, in which disabling differences translate into marginality and exclusion rooted in economic, political and cultural oppression, disabled women need to actively re-place themselves: to create spaces in which marginality and marginalised collective identities can be embraced and valued. Such a re-valuing of disabling differences in turn makes it possible to venture across boundaries of exclusion and to disrupt and challenge ableism in multiple spaces of everyday life. These personal accounts help to illustrate the sorts of life passages that these struggles entail.

The next section of the chapter shifts our attention to the societal scale, in order to illustrate how disability situates women within the political economy and life spaces of late capitalist societies. Using the example of disabled women in Canada, I show how disabling differences help to situate women at the margins of life, both socially and spatially, and to create multiple barriers that help to exclude them from many taken-for-granted spaces of everyday life. I argue that the restructuring of the state is intensifying this exclusion, making it increasingly difficult for disabled women to 'take up ableist spaces' in ways that challenge these women's oppression. At the same time, however, disabled women continue to fight back and, in particular, to create spaces in which they are included and valued. These struggles include passages at the personal scale but also in terms of the collective identities that are negotiated and contested within and between spaces of activism – identities such as those of the women's movement, and of the disabled women's movement.

I conclude with some reflections on how the geographies of disabled women's lives and struggles can be better understood through critical perspectives that take personal and collective rites of passage into account. I also consider the challenges facing disabled women activists in Canada.

Taking up ableist spaces: disabled women's passages

Coming to terms with being constructed as negatively different as a result of disability often involves periods of personal crisis, transition and growth that are both painful and liberating. To illustrate this, I first draw on my personal experiences as a disabled female academic.

On being a disabled woman in academic spaces

Life on the margins of academic decision-making and power is a common experience for many women in the University. This is also true of relatively privileged groups, such as female professors. Despite growing awareness of the barriers created by discrimination on the basis of gender, sexist practices continue to devalue women's presence and contributions in spaces of academic life. As a junior female professor in an all-male department, I experienced cultural oppression on the bases of both gender and academic interests. The former ranged from practices such as sexist remarks which suggested that women belonged someplace 'other' than academia (e.g. in the home), to exclusion from male deliberations regarding issues affecting the department. It included more overtly hostile practices such as portraying female academics as inappropriately ambitious and less competent than their male colleagues. My academic interests in radical geography, in particular in political economy and feminist geography, were controversial, and provided an additional basis on which I (and sometimes my students) were constructed as 'negatively different from others'.

When I became physically disabled (by rheumatoid arthritis) four years after beginning my career as a full-time assistant professor, these early experiences continued to shape my life and to complicate my experiences of cultural oppression on the basis of disabling differences. I did not fully appreciate this at the time, as I have discussed elsewhere (Chouinard 1996). In fact, the gradual realisation of why I experienced my disabling differences the way that I did was an important personal passage which has helped me not only to come to terms with my chronic illness and disabilities, but also to find the determination to contest barriers faced by persons with disabilities in academic spaces. I gradually realised, for example, that my experiences of being devalued as a female academic had helped to discipline me into being less assertive – making me initially reluctant to once again be perceived as a 'difficult woman', in that I demanded that my disability be accommodated, instead of my being forced to give up my academic work and to go on full long-term disability. The latter option was the only one entertained by the University administration during most of the long and often lonely struggle. I had to fight to keep my position, to gain physical access to my office – which was in a non-accessible building – and to negotiate a workload arrangement which would allow me to continue with my academic work without exacerbating my poor health.

When my illness became severe, and joint inflammation and pain made it difficult and sometimes impossible to walk or move much at all, my life was thrown into a phase of personal crisis. Suddenly, I was no longer able to negotiate the spaces of everyday life in 'normal, taken-for-granted' (able-bodied) ways. Barriers of access and understanding seemed to be everywhere. A first, rather rude, encounter to such barriers was a protracted fight for a parking space adjacent to my building, in order that my precious and very limited energy would not be wasted before I even reached my office. Despite repeated statements by my medical specialist that provision of such a space was essential, the director of parking services was not only unsympathetic, but actively tried to undermine my request. What seemed to be a straightforward and rational request was thus turned into a draining and demoralising contest of wills over a relatively small request, and one presumably in the University's interest in terms of facilitating what work I was able to do despite being very ill.

Other battles followed and it was often difficult not to feel overwhelmed. Some senior colleagues reinforced notions that a disabled professor was a 'lesser' professor by telling me, for instance, that someone receiving disability benefits would be 'written off' and no longer considered to be an accomplished member of the academic community. Others stressed the importance of attempting to do all the 'normal duties' of a professor if I hoped to continue to have a position in the University. I experienced a great deal of pressure to continue with all of my undergraduate teaching, for example, even though it was becoming clear that my efforts to do so were resulting in incapacitating flare-ups of the disease, as rheumatoid arthritis is a highly activity-sensitive disease.

Further complicating these experiences was the phenomenon of 'denial' both on the part of myself and of some of my colleagues. Surely, we believed, the next drug treatment would 'work' and make me 'normal' again, and able to take on all the duties of a professor. It took two years of unsuccessful treatment before I accepted that this 'miracle' was probably not going to occur. Some colleagues, however, were not ready to accept this, making it more difficult for me to negotiate a working arrangement that would take my disabilities into account. This, my personal passage into a phase of transition, was out of step with these colleagues' abilities to accept what had happened to me.

I was out of step, also, with those in positions of decision-making power within the University. I learned, for example, that the University had a legal duty to accommodate employees with disabilities. This was made clear to me by a lawyer assigned to my case by the Canadian Association of University Teachers, who found me in tears during our first meeting because I had been told once again that I had either to resume all normal duties of a professor, or to lose my position. When I raised this with my Dean and Chair, neither of whom were familiar with the law regarding disabled employees, I quickly realised that lack of awareness on the part of those in power was another barrier that people like myself had to struggle to overcome on a daily basis. Long before decision-makers within the University were prepared to even consider this notion, I had become convinced, through both my own experiences and discussions with equity experts, that accommodation was a matter of human rights and social justice. Here again, I was out of step.

Today, after five years of struggle and three years as an accommodated academic, the crises and uncertainties of becoming and being a disabled woman professor have given way to a more stable period in which, through measures such as a reduced workload and physical modifications to my building and home, I have been able to carry on with my academic work. That work has included contributing to the development of an accommodation policy for my University covering various special needs (e.g. disability, child care, religion) and promoting the rights of marginalised groups on campus. My own 'test case' has helped to increase awareness of disability and accommodation issues on campus and I have been pleased to find more colleagues recognising the need for action on these issues.

The personal passages I have briefly described here have been difficult ones. This is in part because they inevitably brought me, and those helping me, up against the establishment and against barriers of difference. But it is also because struggles to empower the disabled are waged largely by people who have committed themselves to the struggle for disability rights; and because of this their personal passages are often out of step with those in power. Perhaps this is what struggles to make space for disabled women and men in academia and beyond are all about: *disrupting* spaces of power and privilege to the point where those who dominate such spaces are forced to recognise the difference that disabling differences make, and are confronted with the roles they play in sustaining such cultural oppression. In this way, personal crises and transitions can become the basis for bringing about collective, collaborative passages toward greater social justice in spaces of everyday life.

Passages at the margins: disabled women activists in Canada

How does my experience compare with that of others? Disabled women engaged in struggles to advance their human rights and well-being in Canada have done so from positions which they often describe as extremely marginal. Maria Barile (1993), for example, describes disabled women as an 'exploited underclass' in Canadian society, subject to discrimination on the basis of bodily difference, extreme poverty, social isolation and exclusion, and vulnerable to exploitation in institutional settings such as sheltered workshops. Others have emphasised the multiple forms of cultural oppression experienced by disabled women activists.

From the origins of the disability movement in Canada during the 1970s, disabled women found themselves on the 'outside looking in'. Sexism helped to sustain a movement dominated by disabled men. Efforts to place women's issues on its agenda were dismissed or ignored, and disabled women were usually confined to roles such as carrying coffee or completing committee work for male leaders (Driedger 1993: 175–176). Furthermore, when these and other experiences encouraged disabled women to seek alliances with feminist activists, they quickly discovered that they were relatively 'invisible' sisters in the supposedly common struggle for women's rights. In a moving statement, Pat Israel (1985) described the humiliation and pain she experienced when, at a national women's conference, she discovered that the only way in which she could access events was through using a decrepit and dirty freight elevator:

> Sometimes I feel like crying and screaming and I just want to quit the feminist movement. But I can't … I will always fight for the rights of women. I just wish that women in the women's movement would recognise us as sisters and instead of putting barriers in front of us would open the doors and welcome us. Years ago when I attended a national women's conference I had to use a dirty, foul-smelling elevator to get to the workshops. There was garbage on the floors and walls. I felt degraded and dirty every time I had to use it. I wonder what would

> have happened if the black women attending the same conference had been told to use the freight elevator because they were black.
>
> (Israel 1985: 1–2)

In 1986, an open letter to the women's movement in Toronto, reprinted in several women's magazines and written by the local chapter of the recently established DisAbled Women's Network (DAWN Toronto), protested at the exclusion of disabled women from spaces of feminist organising. Tragically, its authors pointed out, disabled women were so far 'off the map' of the women's movement that feminist leaders did not think to ensure that women's events were accessible to disabled women or even to publicise those events in ways that would reach their disabled sisters:

> Who would think of putting out a flyer saying: IMPORTANT FEMINIST EVENT FEATURING MS. DARING DAIRY, WELL KNOWN AUTHOR, Nov. 30, 8:00 PM, Everywoman's Hall. Admission FREE. Childcare. DISABLED WOMEN NEED NOT APPLY. Of course not! Yet often, even usually, that's what the publicity for feminist events says to disabled women ... Your problem is usually that you just plain don't know what accessibility is. Our problem is that we can't get in to even tell you.
>
> (DAWN Toronto 1986: 80)

Sexual orientation has been another difference shaping disabled women's experiences of and passages through the disability and women's movements. A lesbian disability activist reported experiences of constantly being 'out of place' in struggles for disabled women's rights. In the lesbian community, she noted, it was her disability that made her a different and 'lesser' participant. When she was in spaces of organising around disability issues, on the other hand, her sexual orientation became the key marker of difference and a basis of exclusion (Doucette 1991). Politically, then, there were few safe spaces in which her identity as a disabled lesbian activist could flourish. These experiences prompted her to initiate educational efforts within both movements, with the aim of attempting to raise awareness of disability issues within the women's movement and of differences in sexuality within the disability movement. Such initiatives have been important in allowing disabled women to move more easily between different spaces of political action (Doucette 1991). Sadly, however, they have been insufficient to ensure greater general tolerance of diversity within the women's and disability movements. Concerns about lesbian participation and leadership within disabled women's activism in Canada have, for example, continued to be a source of conflict amongst disabled women. The feminist national organisation DAWN and its regional and local chapters have been flashpoints for such tensions and conflict (Stone 1989; Doucette 1991; Driedger 1993; Odette 1993).

Intolerance to diversity amongst disabled women has had painful personal and political consequences. A disabled lesbian activist reflected in the following way on her exclusion and alienation from the disability movement in Canada:

> I have heard some pretty homophobic personal remarks by disability rights leaders, remarks which were not challenged. I do not particularly want to get involved with such groups, not because I am a separatist, but because I feel wounded by such comments. Some disabled women at DAWN objected to a minority rights clause in our constitution which enshrined lesbian rights. An Open Letter DAWN did to the disability rights movement on homophobia and mailed to dozens of disability rights organisations received absolutely no response – nothing. It was like dropping a pebble into a dark well and hearing nothing, not even a splash. I know some people discourage family members from attending DAWN because there are lesbians here. The homophobia is very real.
>
> (cited in Doucette 1990: 64)

Coping with homophobia as well as other cultural oppressions (e.g. differences in race and ability) often precipitates personal crises for disabled lesbian women activists. Doucette (1990) notes how the disabled lesbian women she interviewed were often forced to deny important aspects of themselves in order to survive. They kept their lesbian orientation 'in the closet' in order to retain jobs, and hid disabilities they had been taught to reject (rather than embrace) as part of themselves. For some of these women, making the transition to disclosing their lesbian or disabled selves was a significant act of personal resistance, validating aspects of their identities they had been encouraged to deny. This disabled lesbian woman described the empowerment associated with her 'coming out':

> About six or seven years ago, I realised, 'That's enough! I am suppressing part of myself' – I am proud [now]. I feel I'm more complete. I'm not hiding a part of myself. When you suppress one part, before you know it, you suppress all of yourself
>
> (cited in Doucette 1990: 65)

Several collective 'rites of passage' can be identified in disabled women's activism in Canada during the 1980s and 1990s. Arguably one of the most important was the establishment of the DAWN in 1985. The development of a separate disabled women's organisation marked a turning point with respect to political action. No longer content to try to work from within established disability and women's organisations, an energetic group of feminist disabled women had decided to make disabled women's issues, such as reproductive rights and domestic violence, 'their own' and to strike out on an independent course of research and organising at national and local scales (Stone and Doucette 1988; Stone 1989; Doucette 1991). DAWN's work has been important in raising awareness of the challenges and issues facing women with disabilities and in nurturing disabled women's leadership and networking. However, many disabled women remain 'outside' this arm of the disabled women's movement, continuing either to work within disability organisations or living under conditions which place them beyond the 'reach' of political organising: in institutional settings, in communities without formal organising networks or confined to their homes due to poverty and declining access to services such as paratransit (government-funded special transportation services for persons with disabilities).

A second important collective rite of passage concerns the production of knowledge about disabled women's lives. Over the years, disabled women's experiences of being subjects of academic research – of being sources of data which help to advance academic careers and expert knowledges of disability issues – has contributed to feelings of exploitation by and alienation from the academic research community. This has been most evident in the case of DAWN which has developed an independent course of feminist research conducted in what are claimed to be more inclusive ways that give voice to disabled women's knowledge and concerns. Like other marginalised peoples in our societies, members of DAWN have expressed frustration at 'being researched'. They have staked a claim by and for disabled women to conduct research in ways that respect disabled women's knowledges and their rights to apply that knowledge in ways that advance their collective well-being (Stone 1989: 131–132; personal communications). Such collective passages have contradictory implications for the empowerment of disabled women activists. On the one hand, insistence on the importance of disabled women having a stake and voice in knowledge about their lives is undoubtedly empowering at a time when 'expertise' continues to be a relatively exclusive domain. On the other hand, DAWN's rejection of academic research cuts off significant opportunities to tap the resources of institutions such as the University and to network with researchers sympathetic to and conversant in more inclusionary research approaches. These opportunities are likely to become increasingly important to research endeavours if, as seems likely, government support for disability organisations continues to decline. In addition, such rifts exacerbate the political isolation of disabled women students and researchers within academic institutions.

Collective passages are also being prompted by current crises in state assistance to disabled individuals and organisations, and within disability rights struggles. In Ontario, cuts by the Conservative government of Mike Harris to an array of services and programmes used by persons with disabilities (including social assistance, subsidies for assistive devices, health care, enforcement of legal protections of rights) has helped to fuel political organising for a provincial Ontarians with Disabilities Act (ODA). As a lawyer involved in the provincial ODA Committee recently indicated, it has become clear that piecemeal reforms of law through test cases have been insufficient to protect the rights of the disabled in Ontario (ODA Committee 1997). Although the Ontario government continues to resist this major legal reform, the need for the ODA has become a rallying point for the disabled in communities across the province.

Political activism has, for disabled women in Canada, been in many ways a matter of grappling with multiple and shifting crises, and of moving within and between multiple spaces of marginality. Disabled women have made significant inroads into their political invisibility, both in the disability and the women's movements. Their ability to sustain a collective political presence has, however, been under attack as cuts in government support exacerbate the problems of mobilising an extremely impoverished and often isolated constituency. Compounding these difficulties are the country's size, diversity of living conditions, and barriers of difference within disabled women's activism. Responding to barriers to collective political action has frequently placed disabled women 'out of place', i.e. in the role of outsiders or of 'lesser members' in movements they believe in. Being marginal everywhere is a demoralising and alienating experience. However, to their credit, disabled women continue to contest spaces of exclusion and to establish voices within organisations such as the National Action Committee on the Status of Women. Building on these strengths in the political activism of the next millenium will involve new crises and transitions. One key challenge is to convince all Canadians that they have a stake in advancing the rights and well-being of disabled women, men and children. Another is to practise inclusion of different women within disabled women's struggles themselves.

Being 'out of place': disabling differences and women's lives

The social construction of disabling differences as markers of inferiority and 'otherness' has, at both macro and micro scales, helped to situate disabled women as 'out of place' in society and in the spaces of everyday life. In this section, I illustrate some important facets of what being 'out of place' means in the Canadian context.

Statistics on employment earnings, unemployment and poverty rates clearly demonstrate the extreme marginalisation of disabled women in the Canadian labour force and economy. In 1997, the Ontario chapter of DAWN estimated that 70 per cent of disabled women of working age were unemployed. This is a staggeringly high rate, especially if compared to a national average unemployment rate in 1997 of 9.2 per cent (Statistics Canada 1998). Disabled women who are employed earn less income than disabled men (who in turn earn considerably less than able-bodied female or male workers). In 1991, disabled women aged 15–34 earned an average of 68.7 per cent of the employment income earned by disabled men in the same age range (who received an average of only 70 per cent of the earnings of non-disabled workers in the same age range). For older disabled women, aged 35–54 and 55–64, average employment income was only 54.7 per cent and 62.6 per cent, respectively, of that earned by disabled men in the same age ranges (Statistics Canada 1995). More disabled women than men live in poverty in Canada: 65 per cent in 1991 compared to 43 per cent of disabled men (Barile 1993). These rates compare to a national poverty rate in 1991 of 16.5 per cent (Canadian Council on Social Development 1998); and such statistics are in many ways only crude indicators of the extent to which disabled women are economic 'outcasts'. To capture the latter one also has to consider aspects of their 'place' within economic life such as: lack of union protection, exploitative and oppressive conditions of work even within environments supposedly adapted to their needs (e.g. sheltered workshops), the inability to purchase commodities many take for granted, and financial and service barriers to accessing multiple social spaces even within the confines of local neighbourhoods.

Culturally, at both national and local scales, disabled women are also 'out of place'. Cultural events, such as theatre, rarely include actors or directors with disabilities and plays which portray or celebrate disabled lives are at least as rare. Popular media, including films and television, also continue to portray successful and attractive women as those who are 'fit' and able-bodied. Disabled male heroes are somewhat more common, for example in films documenting experiences of Vietnam veterans (e.g. *Forrest Gump*). Token efforts at cultural inclusion in many ways seem only to draw attention to the cultural absence of disabled women everywhere else. In local spaces of everyday life, cultural practices contribute to oppressive living environments for disabled women. These practices include unsupportive professional and informal care-takers, invasive questioning by strangers, aversive reactions to the presence of the disabled in public spaces of various types and construction of local spaces which either exclude or segregate disabled users (e.g. government council chambers which lack disabled seating and sign language interpreters; local arenas and movie theatres which provide only segregated seating for wheelchair users).

Disabled women respond in diverse ways to living conditions and social practices that mark them as 'lesser others' who are 'out of place' in Canadian society. As Iris Young (1990) points out, such pervasive economic and cultural oppression necessarily and repeatedly reminds members of oppressed groups that they are different from and devalued by dominant groups in society. Unable to identify with dominant norms, such as the 'able' body, disabled women learn to identify with women who share their experiences of oppression as they seek affirmation of their selves and opportunities for mutual support and action. This in turn can become a basis for collective challenges to oppressive cultural norms. For example, one of the missions of DAWN in Canada has been, and continues to be, to conduct and share research which challenges prevailing images of the idealised and beautiful female body, and various related myths, such as that sexuality is not a part of disabled persons' lives. Other goals include pressing for changes in society that will help to protect and promote the citizenship rights of persons with disabling differences (such as lobbying for a provincial Persons with Disabilities Act in Ontario).

As mentioned above, many disabled women exist on the margins or outside of spaces of collective action. There are many reasons for this. One is that some disabled women internalise negative cultural messages about their self-worth and lives and withdraw from social interaction. Others find it increasingly difficult to overcome multiple barriers to social interaction. These include dwindling access to services such as paratransit and funding cuts to disability organisations which mean that fewer subsidies are available to help cover travel and other expenses related to participation (see Chouinard 1998). Yet another reason, noted above, are the barriers of difference within disabled women's organising such as sexual orientation and race.

State restructuring of programmes assisting disabled individuals and organisations, in industrialised nations such as Canada and the USA, is playing an important role in exacerbating the isolation and socio-spatial marginalisation of disabled women. In Canada, the federal Liberal government is in the process of withdrawing its support for disabled Canadians, both in terms of ending its involvement in disability policy and programmes, and of reducing financial assistance to persons with disabilities. A federal government programme to develop and implement a national strategy for the integration of persons with disabilities within Canadian society has been terminated. Changes to the federal Income Tax Act have made eligibility for the Disability Tax Credit conditional on earned income, leaving approximately two million disabled Canadians ineligible for this type of income assistance (McDonough 1997). Such actions have been coupled with reductions in federal funding to disability organisations. This has made it increasingly difficult for organisations such as DAWN to continue to function; some local chapters have been forced to close and others to make drastic spending cuts. At the national level, funding cuts have made it difficult even to hold regular meetings (since without financial subsidies most disabled women cannot afford travel and accommodation expenses – Chouinard 1998).

Other agencies and programmes assisting the disabled have been casualties of an Ontario government committed to deep cuts in social spending in order to deliver tax cuts to more affluent citizens and promote a 'pro-business' investment climate. There have been reductions in provincial funding for home care services and in the paratransit service across Ontario. Thus, persons with disabilities are likely to be isolated in their homes for longer periods of time and to experience greater difficulty in accessing services and activities within their local communities (McMaster University 1997).

Government funding cuts have also reduced access to legal remedies for the barriers and discrimination faced by disabled persons. Funding to the Ontario Human Rights Commission has been reduced, severely restricting the agency's scope of work. A provincial task force recently recommended that provincial human rights legislation be amended so as to reduce substantially the responsibilities of employers to accommodate employees with disabilities (i.e. to make reasonable modifications to their work spaces and work tasks in order for them to perform the essential duties of their jobs). A telling indicator of the government's lack of commitment to promoting more equitable work places in Ontario is the fact that it abolished a provincial fund specifically devoted to removing systemic barriers to employment within the public service (*Ontario Hansard* 1997). Like the federal government, the Ontario government has also reduced funding for organisations promoting the rights and well-being of the disabled. Local chapters of DAWN have been forced to close. DAWN Ontario has been forced to take drastic actions to continue functioning, including moving from Toronto to a smaller, more remote urban centre in northern Ontario in order to cut costs.

These sorts of changes in state policies and programmes, and resistance to grass-roots initiatives such as the Ontarians with Disabilities Act, are another facet of the multiple ways in which the actions of dominant groups are helping to place disabled women, children and men even further 'outside' Canadian society. Economic and cultural marginalisation is compounded by political visions and agendas in which disabled citizens are largely invisible and often ignored. For disabled women, who are subject to systemic discrimination, both in ableist and gendered forms, it is especially critical, and especially difficult, to contest their location on the margins of political life. Multiple barriers to political action, including extreme poverty, devaluation by potential allies in both the disability and women's movement, and recent trends in state restructuring, have frustrated their passages into political activism.

Contesting boundaries: embracing difference and disrupting spaces of oppression

What is remarkable and inspiring about the personal and collective passages discussed in this paper is that, despite often seemingly overwhelming odds, many disabled women continue to struggle for their places and rights as citizens on the political maps of local, national and international communities. And, as some of the experiences I have discussed indicate, disabled women's experiences of struggles against oppression are punctuated by important phases and moments of crisis, transition and growth – phases and moments which are sometimes personal and at other times shared with others in ways that encourage and shape political action. Geographies of disabled women's lives and struggles that recognise such passages can enrich our understanding of oppression and resistance in a number of ways.

For example, such accounts draw our attention to the significance of personal transition and growth in the development of disabled women's capacities to struggle for social change, both individually and collectively. They also alert us to the fact that struggles to change the environments in which disabled women live, be they universities, work places or other local communities, are not neat, holistic processes but unfold in contradictory ways that, in part, reflect disjunctures in personal and collective passages. One good example of this is disabled women's struggles for accommodation in the work place – struggles which repeatedly bring activists committed to disability rights up against powerful employers and others who often have yet to recognise or accept that they have a legal or moral duty to accommodate disabled employees. Similarly, political divisions within the disabled women's movement in Canada can be understood, again in part, in terms of the different ways women position themselves with regard to debates over such issues as whether or not lesbian women should be included in disabled women's organisations. The passages women go through as they grapple with such issues can be thought of in terms of multiple paths of self-discovery and of relating to different others – multiple because sometimes such paths converge but at other times they may diverge so much that new boundaries and barriers are created within struggles for social change.

At the present juncture, right-wing conservatism and backlashes against efforts to promote greater equity for marginalised groups in Canada have helped to deepen the crisis conditions in which disabled women are contesting their oppressions. It is therefore particularly important that, in addition to drawing creatively on their collective identities, experiences and strengths, disabled women re-visit barriers of difference which have helped to fragment, divide and weaken their political activism. These barriers include differences such as sexual orientation and race, and also differences of social and geographic location. With respect to the latter, the disabled women's movement will grow, mature and flourish only to the extent that its own exclusionary boundaries are disrupted and transgressed – for example, through outreach to disabled women currently isolated from political spaces of struggle; by disabled women activists recognising their own relative privileges (e.g. in education, income and access to political networks) and seeking to empower less fortunate sisters; and through concerted efforts to overcome divisions based on intolerance to diversity amongst disabled women.

These will not be easy transitions or initiatives, particularly when resources are extremely scarce. But when you are struggling against all odds for a collective place and voice in spaces of social life, places from which people with disabling differences can be empowered, you simply cannot afford spaces of resistance that re-draw, rather than challenge, oppressive margins of difference and exclusion. If disabled women are to continue to nurture pro-active collective identities and capacities for struggle over the long-term, they will need to 'subvert the margins from within' through valuing, celebrating and tapping their own diversity. Through collective passages, such as contesting homophobia within spaces of political organising, disabled women can build the strength and determination to continue to fight for spaces of everyday life in which differences such as disability are viewed not as something to be 'corrected' and avoided, but as part of the spectrum of human experiences that enrich all of our journeys through life and society.

References

Barile, M. (1993) 'Disabled women: an exploited underclass', *Canadian Woman Studies* 12, 4: 32–33.

Canadian Council on Social Development (1998) *Poverty Rates, All Persons, Canada 1980–1995*, Canadian Council on Social Development: Centre for International Statistics (available on-line at: http://www.ccsd.ca).

Chouinard, V. (1996) 'Like Alice through the looking glass: accommodation in academia', *Resources for Feminist Research* 24, 3–4: 3–11.

—— (1999) 'Body politics: disabled women's activism in Canada and beyond', in H. Parr and R. Butler (eds) *Mind and Body Spaces: Geographies of Disability, Illness and Impairment*, London and New York: Routledge.

DAWN Toronto (1986) 'An open letter from the DisAbled Women's Network, DAWN Toronto to the women's movement', *Resources for Feminist Research* 15: 80–81.

Doucette, J. (1990) 'Redefining difference: disabled lesbians resist', in S. Dale Stone (ed.) *Lesbians in Canada*, Toronto: Between the Lines Press, 61–72.

—— (1991) 'The DisAbled Women's Network: a fragile success' in J. Wine and J.L. Ristock (eds) *Women and Social Change: Feminist Activism in Canada*, Toronto: Lorimer, 221–235.

Driedger, D. (1993) 'Discovering disabled women's history', in L.E. Carty (ed.) *And Still We Rise*, Toronto: Women's Press, 173–187.

Israel, P. (1985) 'Editorial introduction', *Resources for Feminists* 14, 1: 1–2.

McDonough, A. (1997) 'Equality for persons with disabilities', commentary by Canadian NDP leader (http://www3.sympatico.ca/alexa.halifax/disablefact.html).

McMaster University Summer Gerontological Institute Proceedings (1997) Special session on Advocacy and Disability Research, June, McMaster University: Hamilton, Ontario.

Odette, F. (1993) 'Women with disabilities: the third "sex" – the experience of exclusion in the movement toward equality', unpublished independent inquiry project in partial fulfilment of requirements for the Masters of Social Work Degree, Faculty of Social Work, Carleton University, Ottawa, Canada.

Ontarians with Disabilities Act Committee (1997) *Community Forum*, December, Hamilton, Ontario.

Ontario Hansard (1997) Members' debates on Access for the Disabled, 15 May, available on-line at: http: //www.ontla.on.ca/hansard/ hansard.htm.

Statistics Canada (1995) *A Portrait of Persons with Disabilities*, Ottawa: Statistics Canada.

—— (1998) *CANSIM, Matrix 3472*, Ottawa: Statistics Canada.

Stone, S. Dale (1989) 'Marginal women unite! Organizing the DisAbled Women's Network in Canada', *Journal of Sociology and Social Welfare* 16, 1 127–145.

Stone, S. Dale and Doucette, J. (1988) 'Organizing the marginalized: the DisAbled Women's Network', in F. Cunningham and the Society for Socialist Studies. (eds) *Social Movements/Social Change: The politics and practice of organizing*, Toronto: Between the Lines Press, 81–97.

Young, I. (1990) *Justice and the Politics of Difference*, Princeton, New Jersey: Princeton University Press.

10

JOURNEYING THROUGH M.E.

Identity, the body and women with chronic illness

Pamela Moss and Isabel Dyck (Canada)

My Reality

I lie as if dead
Body glued to bed, like lead
Skin tingling
Muscles twitching
Ears ringing
Numb
Immobile in pain and devastation
Angry at inconsideration
Lack of listening
Neglect to hear
I am not a textbook
This is not a perception
I know these symptoms are real
(Rhoda Howard, 1995)

Metaphors can be useful in coming to terms with the daily lives of women who have been diagnosed with chronic illness. They can assist us in understanding what it is like not only to be healthy, but also to be ill. Yet we must carefully consider the implications of using metaphors. If we do not, then our understanding will be distorted. What we were once interested in becomes an extension of the metaphor, no longer an entity unto itself. In this chapter, we want to work through the implications of applying the metaphors of journey and rites of passage to our understanding of the daily lives of women with chronic illness.

We understand a *rite of passage* to be a ceremonious, celebratory act which publicly marks the movement of an individual from one social category to another, as for example, weddings, confirmations, bar mitzvahs and graduations. In the context of chronic illness, rite of passage can be a useful term, but not necessarily in its conventional sense. Applying a rite of passage as a marking of the transition between illness and health implies a sense of wholeness, in that at one point a person is complete but ill with X and, at another, complete and healthy without X. For women diagnosed with chronic illness, such an implication cannot access the complexity of being ill and being healthy when the state of illness or health fluctuates unpredictably, giving rise, for example, to the common descriptive phrase, 'good days, bad days' (e.g. Charmaz 1991).

When linked with the notion of a journey, the metaphor *rite of passage* begins to make more sense. Using the concept of journey to understand a process, in this case, having chronic illness, implies some sort of direct movement from point A to point B. Such a metaphor of movement alone, however, sets as a norm that the destination of health is the singular goal. For women with chronic illness, this journey is often indirect, sprinkled with what can be considered side trips on rough terrain. As such, the implications of journey may neglect formative experiences throughout the course of a chronic illness, as for example, how going back and forth between being ill and being healthy plays itself out in particular places and how the course of illness influences decisions on what path to take and to which destination. It is here where joining the two metaphors, perhaps phrased as 'journey as a rite of passage', can be useful in coming to terms with the daily lives of women with chronic illness. The journey comprises the experiences of the fluctuations themselves, the inconsistencies of not being able to depend on the physical body and the variable conditions of existence. The rites of passage are specific markers that indicate to the women and their cultural reference groups their illness and how these women weave themselves, as both ill and healthy, into their daily lives. By using the concept of rite of passage in the context of journey, we can think through how a diagnosis can mark a body and how that then shapes the daily life of a woman with chronic illness and launches her journey from a state of health, to one of illness, and, perhaps, back again.

In the first half of this chapter, we provide some background for understanding debates about identity, body and politics, concepts within which to locate a woman's journey. In the second half of the paper, we highlight our arguments by drawing on in-depth interviews with fourteen women diagnosed with *myalgic encephalomyelitis* (M.E.) who live in the Victoria region in British Columbia, Canada.

Identity

Throughout the 1980s and into the 1990s, many feminist geographers focused attention on identity and difference. Two primary streams of feminist thought framed this dialogue: second-wave feminists and feminists informed by post-structuralist thought. The former came to be known as essentialists, the latter, anti-essentialists. This simplistic categorisation highlighted the notion that second-wave feminists used *woman* as the main category of feminist analysis, whereas feminists informed by post-structuralist thought questioned the notion of *woman* itself, in that such a monolithic category could not capture the nuances of all women's experiences, let alone just one of them. This led to conceiving women's identities as multiple, fluid and sometimes place-specific. Further, not only did women have multiple identities manifest in different places at different times, but the identities themselves were fractured, disjointed and variously shaped by numerous sets of power relations in addition to gender. Many second-wave feminists resisted this argument and pointed out how, when taken to extremes, the post-structuralist argument would make women powerless and incapable of action, and had the potential to render women, who are already marginalised by dominant sets of power, voiceless.

As a result of these contrasting perspectives, feminist geographers are now sifting through complex arguments about how much credence to give one position over another. In one sense, invoking difference means dissolving sameness, but, at the same time, recognising that women positioned multiply and variously along several axes of power can thwart attempts at political action grounded in female solidarity. In another sense, when women are united only in the struggle against patriarchy, then the struggles against racism, heterosexism, ethnocentrism and Minority World[1] privilege are compromised. Kobayashi (1997) calls this the paradox of difference and diversity. When querying how difference is socially constructed, the inevitable political contradictions, too, must be addressed.

But identities are not just about who we are; they are also about who we are supposed to be. In our daily lives, examples of attempts to define who we are abound, especially in the context of the body. Our bodies are socially and culturally defined as beautiful, sexed and sexual, through, for example, consumer advertisements for beauty products, religious doctrine about fertility, childbearing and abortion, and legal age restrictions for consensual hetero- and homosexual sex. Although these bodies are not real (even if in film, video and photographs), they do exist discursively as the body beautiful, the religious body and the legal body. *Discourse* consists of ideas, notions, thoughts, and texts that regulate, define, and shape our material existence (q.v. Foucault 1972). Intersections of real and discursive bodies often have concrete implications for identity, particularly because they act as *scripts* for who it is we should be.[2] For example, Bordo (1993) shows how societal values shape women's notions about what it is to be fat and thin and how this can translate into eating disorders such as anorexia nervosa and bulimia.

For understanding the daily lives of women with chronic illness, issues of difference are significant. Mainstream conceptions of ill women define them as *deviant*, as not conforming to specific social norms. Such identities are monolithic, casting a woman and her body as whole (wholly ill or wholly healthy), changing only in one way (from ill to healthy) and consistent (either ill or healthy). However, identities of women with chronic illness are fragmented (both ill and healthy), in flux (sometimes getting more ill, sometimes getting healthier) and contradictory (ill and healthy at the same time). Some women sometimes act 'normal' in the work place for fear they may be 'found out' and lose their jobs (Moss and Dyck 1996). Women with chronic illness have multiple identities that are sometimes place-specific; any one identity is situated in a complex web of unevenly oppressed and privileged power positions. In this sense, post-structural accounts of the social construction of difference assists in understanding the myriad of multiple identities.

Identity politics

However, as feminist critics point out, when appealing to difference as a political organising tool to differentiate individuals and groups, political impasses are inevitable. These identity politics, often derogatorily and sarcastically referred to as 'IP' in organising circles, set the stage for challenging consistent, unwavering goals of politically based groups. When forging common ground is not *the* goal, then gaining political ground against an opposition is. So, in unions, women's and children's labour rights are sometimes overlooked; in women's housing groups, ageing and non-white women's housing needs are sometimes overlooked; and in poverty advocacy groups, prostitutes' and addicts' needs are sometimes overlooked. To create a space for such disregarded needs, people organise around one particular identity. Using an essential, fundamental identity when it is politically savvy to do so, is known as *strategic essentialism* (Spivak 1988, 1990; see also Barrett's (1987) discussion of difference). Yet, as a political strategy, strategic essentialist political positionings can work in more subtle ways, which support the *status quo*. For example, middle income women will remain part of the hegemony reproducing exploitative wage relations for low income women until they actively engage in the struggle against capital, just as Minority World women will continue to benefit from and maintain their privilege until they mobilise against their world's colonisation and domination of the Majority World.

Increasingly, then, it becomes clear that these identities may not be the most astute departure point for political action because they are not effective in crucial ways. First, they diffuse political strength and groups end up fighting amongst themselves rather than challenging, for example, the domination of insurance companies in designating benefits for particular illnesses. Second, they do not capture the multiple, shifting and fluctuating character of any one person or any one group, and, as a result, for example, efforts to challenge the dominance of biomedicine with alternative medicine and treatment are dampened.

With regard to understanding the daily lives of women with chronic illness, identity politics pose several difficulties. A collective and monolithic identity for women with chronic illness is problematic for there are many types of chronic illness. Because of the dominance of the biomedical diagnosis and biomedicine as an institution, political organising for chronic illness is usually done by type of disease. Ill identities are also not so neatly articulated through sets of power relations, a basis upon which organising is accomplished. Some women with chronic illness are more interested in securing income or maintaining social networks than challenging the existing organisation of the delivery of services for chronically ill people in more isolated areas, especially when support services are readily available for treatment of the illness (Moss 1997). We reckon, too, that organising along the lines of a particular illness is even more riddled with problems when the chronic illness has less legitimacy, as for example, chronic fatigue syndrome and environmental (sensitivities) diseases, in contrast to more biomedically legitimate illnesses such as diabetes, arthritis, heart disease and lupus. Contrary to other identities, which are mobilised in response to oppression, ill identities are not willingly or easily taken on. They are, for the most part, imposed, making the (reluctant) reconstitution of a woman's identity deliberate and painstaking.

Body

Identity and identity politics manifest themselves through bodies, the material ones through which we exist. For some time now, feminist geographers have had an interest in the body, whether it be the ageing, discursive, ill or sexual body (e.g. Bell and Valentine 1995; Duncan 1996; Moss and Dyck 1996; Laws 1997).

Shilling (1993) suggests that there are two types of social theories of the body: naturalistic and social constructionist. *Naturalistic* approaches conceive the body as pre-social, as a biological entity, upon which society is layered (e.g. essentialist feminisms, sociobiology). *Social constructionist* approaches understand the body to be a result of interactions within society, either it is shaped, constrained or invented by society (e.g. discourse analysis, structuration). As a way to resolve the tension arising from these two views of the body, Shilling proposes to use the strengths of both in order to provide a fuller picture of what the body actually is. To make the argument, he draws on Elias' (1939) work on the civilised body and his own work on death. Rather than reiterating their arguments, we want simply to point out that these are studies of *transition*. Shilling focuses on the movement from life into death and all that is associated with that movement, as for example, people's emotions around death, existential crises and technologies that prolong life. By utilising the notion of a socially constructed body, Shilling (1993: 200 and 175–196) says theorists can demonstrate the unfinished nature of the body that is shaped by life choices. And, similarly, by treating the body as biological matter, the same theorist can account for technological intervention into the state of bodily existence (Shilling 1993: 200).

Most feminist geographers do not advocate such *theoretical eclecticism*, where part of one theory is used to explain one part of the phenomenon under investigation and another theory, another part, even though in other disciplines eclecticism is somewhat common (e.g. Wolff 1995).[3] They prefer to conceive such issues as paradoxes (see Rose 1993). We do not think that either proposal is appropriate; we think that they both show us the need to focus on the space that inspires the paradox and the empirical situation that breeds eclecticism. It is in this sense that the notion of transition is important, especially in understanding the daily lives of women with chronic illness. Focusing on the journey, the process of change, the pulsating waves of the sensuality of illness, permits us to grasp the life-world (after Buttimer 1976) of women diagnosed with chronic illness. When ill with chronic illness, women have a heightened sensitivity to the elements that both define who they are and displace their permanence. What 'is' quickly becomes what 'was'; and the 'I' and 'me' they thought they knew so well vanishes. The intense interaction with health practitioners, support group networks and institutions in place to provide and assist in securing financial support, forces the woman to reconstruct her 'self' by lacing together remnants of who she thinks she 'is' and 'was' with an ever-changing, unsteady, unreliable and unpredictable disease process. In order for us to access the everyday, the mundane, the 'stuff' that makes up the daily lives of women with chronic illness, we can conceive the illness as a journey, the rites of passage as public markers of illness and the women as sojourners gaining wisdom through their illness.

Premises for a radical body politics

We call the framework we propose a *radical body politics*. Our framework is radical in that we scrutinise the basis upon which we come to know our social and physical environments and we question the surface meanings of images, texts and actions. We focus on body, indeed bodies, as both discursive and material entities through which we experience those environments. We describe our framework as a politics because we use our interest in the way power is wielded as an entry point into a more extensive understanding and better explanation of the process or phenomenon we are investigating.

For us, then, in the context of a discussion about being ill, being healthy, identity, and identity politics, it is important to conceive the body as both real and figurative, both material and discursive, both sensual and textual. Materially, our bodies circumscribe our existences. In this sense we are sensual beings, ones that feel tactilely, emotionally and sensorially. With chronic illness, women feel pain, fatigue, disorientation, malaise, frustration, alienation, isolation. At the same time, our bodies carry cultural markers that tag us as aged, racialised, sexed, classed, sexualised, disabled or ill. A *cultural marker* is similar to both an artefact and an icon; it is a thing or an object associated with a person's body that gives meaning to a particular group of people who hold similar overarching values. These cultural markers are for the most part visible and reveal to our cultural reference group an aspect of our identity. For example, the combination of a hoarse voice, sweaty forehead and a glazed look, as bodily signs, indicate illness.

For women with chronic illness, however, these cultural markers on the body are rarer and often times invisible. Women with chronic illness do not necessarily use wheelchairs, or employ (publicly at least) any assistive devices. These women 'pass' as 'normal' and healthy in everyday life, the illness hidden away, rendered imperceptible. Even the most crucial cultural marker for women with chronic illness is invisible: the diagnosis.

Contradictorily, this invisibility of illness, while hiding the collective identity of women with chronic illness and dividing women against one another, can also protect the woman from socially dominant definitions of deficiency and deviance, as a woman and as a body. Given the invisibility of most chronic illness, collective identities are sometimes difficult to build. And, as in other political groups, some women are marginalised because of who they are and how they are positioned along multiple axes of power. Ironically, these women are challenged on the same criteria used to define the group as a whole as ill or deviant, as for example, in being 'not sick', 'not sick enough' or 'not sick with [fill in the blank]'. Again, this probably takes place more often with biomedically less legitimate illnesses.

Three major premises underlie our framework for understanding the daily lives of women with chronic illness in the transitional space within a paradox. First, individuals have multiple identities and they manifest differently at different times in different places. Second, identities themselves fluctuate. Any one identity is mobile and impermanent, and coalesces around the circumstances of a particular setting. Third, individuals hold contradictory identities simultaneously. There is no merging of opposites, no oscillation between polarities.

Our chronic illness study

We conducted in-depth interviews with fourteen women diagnosed with *myalgic encephalomyelitis* (M.E.) in the Victoria region of British Columbia, Canada.[4] M.E. is a disease of the central nervous system with no single, known origin. Primary symptoms include debilitating fatigue, pain and cognitive impairment. Other symptoms include, for example, noise- and photosensitivity, anxiety, sore throat, nausea, imbalance, dizziness, muscle weakness, migraine-like headaches, sleep disturbance, forgetfulness and blurred vision (Schaefer 1995; Komaroff and Fagioli

1996). A diagnosis of M.E. is usually one of exclusion, that is, ruling out diseases that have definitive laboratory tests and known organic and psychiatric causes. The length of time the diagnosis takes varies. Although symptoms must persist for at least six months before diagnosing M.E., many women wait several years before obtaining a diagnosis. Such a time lapse is considerably distressing given that suffering without knowing what is wrong further enhances symptoms of anxiety and fatigue. Yet even when diagnosed, M.E. as a disease remains more in the realm of the unknown. Often times what little one can expect from M.E. contradicts a woman's bodily sensations and causes her to doubt her own perceptiveness. Precisely because of the uncertainty of the course of the disease, the script for M.E. is ambiguous, which makes M.E. different from other chronic illnesses that have more complete, conventional scripts, as for example, heart disease or diabetes.

We chose M.E. for the study for two primary reasons. First, M.E. is a marginalised chronic illness, and, as such, would provide information on how women experience illness that is not commonly accepted, biomedically or socially. M.E., popularly known as chronic fatigue syndrome (CFS) used to be called 'yuppie flu' in the mid-1980s. It was only as recent as 1994 that the Centre for Disease Control in Atlanta designated M.E. as an illness. Second, the struggle for legitimacy for women with M.E. is similar to the historical psychological battles over definitions of neurasthenia and hysteria (see e.g. Abbey and Garfinkel 1991; Showalter 1997). By studying the process through which certain illness is legitimated, new light can be shed on how and why women's experiences of illness in particular are devalued and how women implicate this struggle into their daily life.[5]

Each interview lasted approximately one and a half hours. We organised the interview into three parts. First, the woman recounted the process she went through to obtain a diagnosis of M.E. Second, we talked about what impact the illness had on the woman's labour in the home and in the work place, both before and after diagnosis. Third, we discussed more generally the expectations the woman thought others had of her and those she had of herself (see Table 10.1). Note that each woman's name is a pseudonym in order to provide anonymity and confidentiality.

The quest for a cultural marker

For these women diagnosed with M.E., cultural markers play a significant role in communication with their reference groups, for example, family, friends, peers and people with whom they come into contact through daily activities. Such cultural markers have multiple meanings attached to them eliciting, often times competing, values. For example, a diagnosis of M.E. signifies illness. Yet each cultural reference group attaches its own meaning and value to that illness. For

Table 10.1 Socio-demographic profiles of women diagnosed with M.E. with year of diagnosis, Victoria Region, British Columbia, Canada

Name	*Age at diagnosis*	*Present individual annual income*	*Present household annual income*	*Employment status*	*Profession or employment*	*Year of diagnosis*
Agnes	71	$26,500	$40,000	Not employed	Counsellor	1993
Caron	54	$26,400	$86,400	Not employed	Health administrator	1993
Connie	60	$19,920	$35,520	Not employed	Support staff worker	1995
Dolores	44	$24,000	Same as individual	Not employed	Nurse	1995
Elise	44	$60,000	$80,000	Employed	Counsellor	1991
Erin	38	$8,220	Same as individual	Not employed	Floor sales clerk	1996[a]
Jayne	53	$37,680	Same as individual	Not employed	Teacher	1993
Raquel	54	$33,600	Same as individual	Not employed	Project manager	1994
Reann	33	$22,560	$39,360	Not employed	Technician	1993
Sandra	Late 30s	$14,950	Not reported	Employed no sick leave	Healthcare provider	1995
Sophie	61	$36,780	Same as individual	Not employed	Medical technologist	1994
Teresa	49	$8,220	Same as individual	Not employed	Nurse's aide	1996[b]
Verna	Early 50s	Not reported	Not reported	Employed	Community organiser	1993
Vivienne	40	Fluctuating	Fluctuating	Employed	Student[c]	1993[d]

Notes:

a Diagnosed with *fibromyalgia* (FM) in 1995. FM, known formerly as fibrositis and more recently as chronic pain syndrome, is a chronic illness that can be debilitating and is often linked with M.E. FM arises out of a sleep disorder resulting in a set of symptoms similar to those of M.E. People often compare the two by saying that with FM there is more pain and with M.E., more fatigue

b Diagnosed with FM in 1993

c We consider being a student as being employed

d Diagnosed with EBV (Epstein–Barr Virus), a virus historically linked to M.E., in 1987

the woman diagnosed with M.E., it may mean relief, legitimacy and grief. For her employer, it may mean trendiness, hysteria and weakness.

A diagnosis as a cultural marker is crucial for women to negotiate the journey through chronic illness; it acts as the ticket for the journey. Obtaining that ticket is itself a rite of passage, a crossing of a threshold. For each of the women in the study, the diagnosis process was a horrendous experience. Several women were diagnosed with other illnesses prior to being labelled with M.E., such as, multiple sclerosis, fibromyalgia, myelodisplasia, depression, stress and arthritis. The uncertainty in not knowing what caused the pain, fatigue and fuzzy thinking was a source of apprehension and anguish for the women. Often, the proposed treatment for another illness exacerbated the symptoms of M.E., which made for even more discomfort and unease. The length of the process of diagnosis for the women in this study ranged from six months to fourteen years.

Yet, once tagged with M.E., the diagnosis seemed to make sense to the women who were struggling to find out what was making them ill. For those women who knew nothing about M.E., the diagnosis became their script, the primary factor through which they would make sense of their experience of illness. Many of the women viewed our interview as a cathartic exercise through which they could finally get their story told, for many had been informed by biomedical physicians that their pain existed only in their heads and that they could be in paid employment if they *really* wanted to. This rite of passage was fraught with frustration, indignity, sorrow and pain. Yet the will to obtain this particular cultural marker was strong, for with it, a woman could begin accumulating information about the illness and get on with her treatment, but without it, she would have continued to exist in a state of not knowing, a liminal state, which, as the women pointed out, was much worse.

Different groups of people made various uses of this cultural marker. For insurance investigators, a diagnosis of M.E. on a long-term disability claim signalled problems. Because the recovery rate is indefinite, physicians employed by the insurance companies seek to find more 'treatable' diagnoses, for example, depression. Claims with diagnoses of M.E. are scrutinised closely. Only seven of the fourteen women in the study received benefits from private long-term disability insurance companies (see Table 10.2).[6] It is also interesting to note that the women with higher incomes prior to illness were the ones securing both the state pension and long-term disability insurance benefits. Also, the women with less education were less likely to secure disability benefits of any kind. Having M.E. means drastically reduced income if paid employment is not possible.

The women diagnosed with M.E. used the cultural marker as a way to identify themselves as a group. Most of the women in the study told us how important it was to talk only to those who had 'officially' been diagnosed with M.E. The 'invisible' diagnosis became *the* cultural marker around and through which most of these women identified themselves to us and each other. Granted, this was indeed why we contacted them initially. However, the insistence of the women to have us *only* talk to women diagnosed through the biomedical system as having M.E. showed us that they were trying to understand themselves in the context of other women who were experiencing the *same* illness, who had traversed the same rite of passage.

Table 10.2 Descriptions of employment and income

Name	*Age*	*Education*	*Individual annual income*	*Source(s) of Income*	*Annual income before diagnosis*
Agnes	71	Trade school; BSW, MS	$26,500	Old age pension[a]	~$60,000
Caron	54	BA, MSW	$26,400	State disability pension LTD benefits	~$65,000
Connie	60	Trade school	$19,920	State disability pension	~$35,000
Dolores	44	College, Nursing	$24,000	State disability pension LTD benefits	~ same
Elise	44	BSW, MSW	$60,000	Employed as counsellor	~ same[b]
Erin	38	Grade 12	$8,220	Social assistance	~$12,000
Jayne	53	BA, BEd	$37,680	State disability pension LTD benefits	~$60,000
Raquel	54	College, Interior design	$33,600	State disability pension LTD benefits	~$90,000
Reann	33	BSc, MSc	$22,560	State disability pension LTD benefits	~ same
Sandra	Late 30s	College, Dental hygiene	$14,950	Employed as healthcare provider	~$16,800[c]
Sophie	61	Trade school	$36,780	State disability pension LTD benefits	~ $65,000
Teresa	49	Trade school Nursing assistant	$8,220	Social assistance	~$20,000
Verna	Early 50s	BA	Not reported	Employed as Community Consultant	Not reported
Vivienne	40	BA	Fluctuating	Student Research Contract Work	$45,000[d]

Notes:

a Agnes retired just before she was diagnosed with M.E.

b Elise's workload changed drastically. She cut back her private practice and took on a part-time counselling position in an institution

c Sandra went to three-quarters time and her boss gave her a raise to make up for lost wages

d At the time she was ill, Verna received long-term disability insurance benefits. She was working part-time when interviewed

e Prior to her diagnosis, Vivienne was in banking

Passing through this rite to obtain a diagnostic label in order to be accepted as having M.E., either to secure an income or be part of a group, does not mean that the journey has to be the same for everyone. Nevertheless, the women clung to the notion of a collective identity. It was important for the women to have gone through the diagnosis process for M.E. in order to *belong* to the group, in order to know and understand the implications of the M.E. script. To facilitate the construction of a collective identity, they used their social networks, particularly the local support groups founded to provide information to sufferers of M.E. Manoeuvring through the maze of physicians, specialists, state agencies and insurance companies is a central concern of the group. Members informally meet and pass along crucial information, from names of physicians who 'believe' in M.E. to copies of successful long-term disability claims. This tactic of separation between formal and informal networking is politically strategic (though not essentialist because it does not base the act on one fixed identity) in that the group does not contravene any law while maintaining the types of support women need when dealing with M.E., especially initially. Securing income, through state agencies and insurance companies, and obtaining a biomedically sound diagnosis for these forms are the two most important activities that the women engage in initially and that bring them to the support group.

After struggling so hard to get the cultural marker, the same marker can distance a woman socially from her family, friends and co-workers. Her identity around illness fractures her identity as a mother, daughter, friend, lover, wife, etc. Although some women with chronic illness and disability report intimate relationships breaking up once becoming ill, none of the women's life-partners nor any of the women broke off this relationship. Several women did report incidents of loss however. For example, Jayne lost a very close friend once she was diagnosed with M.E.[6] Her friend refused to accept Jayne's 'ill' identity, the one she reconfigured to take into account illness, possibly because her friendship was based not on emotion but on shared activity, now no longer possible.

For Connie, the diagnosis of M.E. contradicted her experience of being ill with breast cancer. She received tremendous support from her co-workers and boss when she had a double mastectomy. However, with the diagnosis of M.E., the same co-workers and boss ostracised her and provided no support. Not only did she feel let down, but she also began doubting her symptoms of fatigue and fuzzy thinking because no one in her previous work environment, all of whom she respected, believed she was ill. She compares one set of meanings attached to illness with the other: she feels she failed at being ill with M.E., because with cancer she was more 'acceptably' ill.

Not being able to plan to socialise is terribly unsettling. Teresa rarely goes out because she is afraid that she will forget the way home or be somewhere when she desperately needs to lie down but can not. Caron misses being able to go out with friends in the evenings for she never knows if she will be able to stay awake for a movie or sit comfortably through a concert. Yet Teresa and Caron take a more pragmatic approach to dealing with the impact the diagnosis has on their social life. They described this externally imposed lack of social interaction as more of a closing off of social contact for healing rather than a shake-down of who really were true friends.

These contradictory experiences amongst the women and within one's own set of experiences emphasise the variability of carrying M.E. as a cultural marker. With chronic illness, diagnosis as a rite of passage marking the body as ill is never complete. With each fluctuating experience of being ill, being told there is no illness, being denied benefits, a woman in some way integrates this information, creating not just a singularly defined ill identity, but multiple identities laced with illness. As time went by, the women, for the most part, became more and more comfortable with these unfixed notions of self.

Putting together a treatment regime

Once a woman has her ticket (the diagnosis), she undertakes her journey of chronic illness, and faces its unpredictability, its uncertainty, its instability. One of the areas where the women were able to apply the underlying tenets of an unfixed, fluctuating identity was in the treatment of M.E. There is no standard treatment for M.E. The women, through advice, research and networking, create their own treatment regimes. Parallel to the way the women were forced into a ceaseless undulation of fixing, unfixing and re-fixing identities, they more consciously used the same technique to create their own treatment regime. Most of the women travelled freely between three approaches to treatment: biomedicine, alternative health therapies and spirituality, each of which is based on a different body of knowledge.

Biomedicine systematically treats the *individual* symptoms of M.E. If the woman is anxious, a physician prescribes anti-anxiety drugs; if she cannot sleep, antidepressants; if she has pain, painkillers. Alternative health therapies are primarily based on treating the person *as a whole*. Some therapies realign the life energy, *chi*, as for example, acupuncture and acupressure; some promote relaxation, posture and overall health, as for example, massage, reflexology and chiropractics; and some use herbs and essences of plants to repair damage to the body, as for example, Chinese herb therapy, homeopathy and Bach flower remedies. Spirituality is based on maintaining positive spiritual energy for a healthy mind, body and soul. Colour therapy, visualisation and praying are all forms of spiritual healing.

Agnes was the only woman in the study who treated her M.E. solely through biomedicine. She reasoned that physicians, although they did not know everything, knew better than anyone else what to do. None of the fourteen women relied solely on alternative health therapies. Many of the women found out about alternative health therapies only after diagnosis, through the support group. Teresa and Erin would have liked to use more alternative therapies, but costs were prohibitive. Even user fees for massage and physiotherapy were too high if used on a regular basis, and, in treating M.E., regular visits are the only way that massage and physiotherapy work.[7] Six of the women placed their spirituality at the centre of their treatment regimes. In fact, Dolores relied heavily on the exploration of her spiritual awareness with a guide to find a healing space for M.E.

In a somewhat eclectic manner, often being more pragmatic than intellectually consistent with the bodies of knowledge they tap, most of the women combined treatments. Like Sandra, Sophie believes in both biomedicine and alternative health therapy knowledges. She sees a psychiatrist and a counsellor, and takes multiple pharmaceutical drugs and several types of dietary supplements, including for example malic acid, grape seed extract and CoQ10 enzymes. In addition, Sophie uses prayer to help sustain her. Connie, too, takes several types of prescription drugs, supplements and depends on her spirituality to help maintain her strength for survival.

After her diagnosis, Reann became a vegetarian and now eats only organic foods. She pays close attention to her body's systems and uses herbal therapies to facilitate their proper functioning. She draws on biomedicine by taking anti-depressants for sleep and an occasional painkiller. Raquel has tried a number of things to assist her in treating M.E. After fourteen years of pain, despair and several surgeries, she now lives with daily intakes of morphine and anti-depressants. She also engages in co-counselling, from which she says she has had immeasurable benefit in coping with both the indeterminacy and unpredictability of M.E. What she finds most gratifying, however, are the times she can spend cooking, sewing, painting and writing.

Like a number of the other women, Jayne draws on all three types of knowledge for her treatment. She sees a psychiatrist, takes anti-depressants and has two breadboxes full of herbs, vitamins and minerals. She also has engaged in or used colonics, silver drops, neurotherapy, meditation, yoga, hypnosis, body work, meditation, cold water baths, Reiki, detoxification baths, chiropractics and fasting. However, she finds what she calls her soul work the most beneficial for healing. Her soul work consists of journalling, reflecting and writing about her journey, about the things she's learned, about the shifts in her life she has made because of her illness.

By combining different approaches to treatment, those based on competing and contradictory knowledge systems, to create a treatment regime for M.E., the women are forging 'healing' spaces that are both comfortable and compliant with their needs. These 'healing' spaces comprise the multiple identities these women have, each with its own physical, emotional and spiritual dimension. The ease with which these women hold these contradictory knowledges show how they are able to develop ways to detach and re-attach their identities to pre-given bodies of knowledge strategically, drawing on one for diagnosis and the treatment of specific symptoms, drawing on another for both the treatment of particular symptoms and developing a more holistic health regime.

From less health to more illness to more health to less illness to . . .

Through the metaphors of journey and rites of passage, we have stitched a tapestry of some aspects of the daily lives of women with chronic illness. We have used a set of key concepts to portray certain elements of these women's lives. We showed how, rather than a single 'ill' identity, these women had multiple and fractured identities, laced with illness. We demonstrated how these women used their identities and bodies in their quest for a diagnosis, or a cultural marker, and in their creation of a treatment regime.

A journey based on a disease process that can force you to be inactive for weeks and be followed by days relatively symptom-free, can toss a woman into a state of disbelief and disorder – a destabilisation of identity. These experiences lead the women to hold simultaneously an ill and healthy, fixed and unfixed, stable and unstable identity. The contradictions they experience through these transitions, through these rites of passage, both collectively and individually, emphasise the diversity of women with M.E. as well as their diverse struggles over illness as a fixed, social identity of deviance.

Assuming a woman, before she is ill, is in a state of health, is probably not a good place to begin any journey into health and illness. Although a diagnosis is the ticket, illness is not only a label, just as it is not only bodily sensations. A radical body politics has to look beyond the surface, into the space of materiality and discourse, in order to understand the complexity of journeys of transition. Just like each undulating symptom of fatigue, pain and cognitive dysfunction, each diagnosis, decision for treatment and visit to a health practitioner, too, comprise the various cultural markers women with chronic illness experience on their journey through less health to more illness to more health to less illness to ...

Transformation

I plummet into the depths of despair
Into blackness, purple rage
Scraping dry blood from inner reaches

Residue of vulnerability, pain, grief
It is time to be over
To become transformed of any security
Dashing the inner system on rocks of the path
(Rhoda Howard, 1993)

Acknowledgements

Thank you, Rhoda, for passing your poems about M.E. to us. Thanks to Blakie Baber for comments on an earlier draft and to Liz Teather for her editing skills. The Social Science and Humanities Research Council of Canada funded this project (No. 410-95-0267).

Notes

1 Like Roth (2000) I use 'Minority' and 'Majority' Worlds in place of phrases like 'First' and 'Third' Worlds, 'the North and 'the South', or 'Global South'. Minority and Majority Worlds are not bounded geographically; rather they refer to colonising relations and economic exploitation within global capital.
2 Cream (1995) discusses the implications of taking the 'pill' had for different groups of women. This shows how one script can dominate a particular act while erasing the multiple experiences of women engaged in the same act, that is, taking the 'pill'.
3 See Fincher (1983) for a discussion of the problems with theoretical eclecticism.
4 For the entire project, we conducted forty-nine interviews, twenty-five with women who were diagnosed with M.E., and twenty-four with women who had a debilitating type of arthritis. Kathleen Gabelmann, graduate student and research assistant, assisted in conducting interviews for the project. For this paper, we use only the information on the women from the Victoria region for two reasons. First, empirically, these women form a more cohesive group than either the other M.E. group in Vancouver or the arthritis group, socially, spatially and politically. In this sense, their transition into a state of illness with M.E. was mediated by many of the same factors, the same script! Second, theoretically, these women's stories most effectively illustrate the concepts developed in this paper.
5 For the comparative aspect of the overall research project, we chose to juxtapose the relatively invisible M.E. with debilitating types of arthritis, which are starkly visible. Comparing such illnesses should be able to provide insight into how women structure and restructure their social and physical environments.
6 The distinction between falling ill and obtaining a diagnosis is important. When the women fell ill with M.E., many still tried to carry on with their regular activity level. Once there was affirmation from a physician that the woman was ill, the woman felt released to begin treating herself as if she were ill and restructuring her daily life.
7 Only part of the cost for physiotherapy, chiropractics and massage therapy are covered by the British Columbia health plan.

References

Abbey, S.E. and Garfinkel, P.E. (1991) 'Neurasthenia and chronic fatigue syndrome: the role of culture in the making of a diagnosis', *American Journal of Psychiatry* 148: 1638–1646.

Barrett, M. (1987) 'The concept of difference', *Feminist Review* 26: 29–41.

Bell, D. and Valentine, G. (eds) (1995) *Mapping Desire*, London: Routledge.

Bordo, S. (1993) *Unbearable Weight: Feminism, Western Culture and the Body*, Berkeley: University of California Press.

Buttimer, A. (1976) 'Grasping the dynamism of the lifeworld', *Annals of the Association of American Geographers* 66: 277–292.

Charmaz, K. (1991) *Good Days, Bad Days: The Self in Chronic Illness and Time*, Berkeley: University of California Press.

Cream, J. (1995) 'Women on trial: a private pillory?', in S. Pile and N. Thrift (eds) *Mapping the Subject: Geographies of Cultural Transformation*, London: Routledge, 158–169.

Duncan, N. (ed.) (1996) *Body/Space*, New York: Routledge.

Dyck, I. (1995) 'Hidden geographies: the changing lifeworlds of women with disabilities', *Social Science in Medicine* 40, 3: 307–320.

Elias, N. (1939) *The Civilising Process, Vol. I: The History of Manners*, reprint 1978, Oxford: Blackwell.

Fincher, R. (1983) 'The inconsistency of eclecticism', *Environment and Planning A* 15: 607–622.

Foucault, M. (1972) *The Archaeology of Knowledge*, London: Tavistock.

Howard, R. (1993, 1995) unpublished poems, made available by the author.

Kobayashi, A. (1997) 'The paradox of difference and diversity (or, why the threshold keeps moving)', in J.P. Jones, H.J. Nast and S.M. Roberts (eds) *Thresholds in Feminist Geography: Difference, Methodology, Representation*, Lanham, Maryland: Rowman and Littlefield, 3–9.

Komaroff, A.L. and Fagioli, L. (1996) 'Medical assessment of fatigue and chronic fatigue syndrome', in M.A. Dematrack and S.E. Abbey (eds) *Chronic Fatigue Syndrome: An Integrative Approach to Evaluation and Treatment*, New York: Guilford, 154–180.

Laws, G. (1997) 'Women's life courses, spatial mobility, and state politics', in J.P. Jones, H.J. Nast and S.M. Roberts (eds) *Thresholds in Feminist Geography: Difference, Methodology, Representation*, Lanham, Maryland: Rowman and Littlefield, 47–64.

Moss, P. (1997) 'Negotiating spaces in home environments: older women living with arthritis', *Social Science and Medicine* 45, 1: 23–33.

Moss, P. and Dyck, I. (1996) 'Inquiry into body and environment: women, work and chronic illness', *Environment and Planning D: Society and Space* 14, 6: 737–753.

Rose, G. (1993) *Feminism and Geography*, London: Routledge.

Roth, R. (2000) 'A self-reflective exploration into development research', in P. Moss (ed.) *Engaging Autobiography: Writing Lives in Geography*, Syracuse, New York: University of Syracuse Press.

Schaefer, K.M. (1995) 'Sleep disturbances and fatigue in women with fibromyalgia and Chronic Fatigue Syndrome', *Journal of Obstetric, Gynecologic, and Neonatal Nursing* 24, 3: 229–233.

Shilling, C. (1993) *The Body and Social Theory*, Newbury Park, California: Sage.

Showalter, E. (1997) *Hystories: Hysterical Epidemics and Modern Media*, New York: Columbia University Press.

Spivak, Gayatri Chakravorty (1988) 'Subaltern studies: deconstructing historiography', in *In Other Worlds: Essays in Cultural Politics*, New York: Routledge.
—— (1990) 'Strategy, identity, writing', in S. Harasym (ed.) *The Post-colonial Critic: Interviews, Strategies, Dialogues*, New York: Routledge, 197–221.
Wolff, Janet (1995) *Resident Alien: Feminist Cultural Criticism*, London: Polity Press.

11

IDENTITY AND HOME IN THE MIGRATORY EXPERIENCE OF RECENT HONG KONG CHINESE-CANADIAN MIGRANTS

Wendy W.Y. Chan (Hong Kong)

> The tender soul has fixed his love on one spot in the world; the strong man has extended his love to all places; the perfect man has extinguished his.
>
> (Hugo of St. Victor, quoted in Said 1983: 7)

Human migration is an age-old phenomenon, although the level and direction of movement have always been dictated by economic factors and government policies at home and abroad.[1] In terms of Chinese migration, Wang (1991) identifies four groups of Chinese migrants. The first, *huashang*, Chinese traders, left China for other Asian countries as early as the Sung dynasty (960-1279). The second, *huagong*, Chinese labourers, left in the later nineteenth and early twentieth centuries, mainly for North America, Australia and New Zealand, some becoming gold miners. A third group, *huayi*, consists of Chinese with non-Chinese nationality such as the Chinese-Canadians I discuss here. A fourth group, *huaqiao*, sojourners, remain overseas only temporarily. In the mass emigration from Hong Kong in the early 1990s,[2] many emigrants cited the reversion of sovereignty to China in 1997 as the main reason for their departure, mentioning concerns about what the future might hold for civil rights, individual liberty, the legal system and living standards (Wong 1992).

Centuries of Chinese emigration have resulted in Chinese living in almost every corner of the world, and there is no lack of literature on the experience of Chinese immigrants living overseas in general including the experience of those living in Canada, the subjects of my study.[3] But there are few studies on the current high return migration to Hong Kong.[4] Thus, the research that I carried out in 1991 and 1994–96 is a valuable addition to the migration literature. It began with my own journey as an emigrant to Canada and was further developed after my return to Hong Kong ten years later.

Emigrating to Toronto

My first study in 1991 explored the impact of migration on the life-style of Hong Kong immigrant women in Toronto (Chan Tang 1991). Ten women, between 25 and 45 years of age, were selected for in-depth interview through the snowball method.[5] The jobs they held prior to their emigration ranged from secretarial/administrative to professional positions. Only two were housewives when interviewed. Five came from Scarborough, a metropolitan area well known for its concentration of Chinese and other Asian immigrants. Five came from other parts of the metropolis.

Using the time-use budget analysis suggested by Michelson and Reed (1975), my study showed that all my interviewees experienced changes in life-style that inevitably reflected the establishment of a different social positioning from that in Hong Kong. In the majority of cases, these changes meant more work, less out-of-home entertainment and no luxuries. The most common reasons cited for these changes were more housework, fewer relatives and friends, longer travelling time and less extra money to spend. Both the productive and reproductive spheres in Canada were different from their counterparts in Hong Kong where the respondents were concerned. The loss of domestic support, provided for many professional families in Hong Kong by full time, live-in Filipina helpers (maids), led many of my interviewees to lower their career expectations.

It is clear that migration had forced many of the immigrant women whom I interviewed to reassess their aspirations and self-concepts – aspects of their very identity. Some of them were far from uncritical of their Canadian context, especially in terms of lost opportunities to fulfil their career aspirations. Lily, a successful businesswoman before she left Hong Kong, commented:

> Sure, I like Canada. The weather is nice though a bit cold; space is good; air is fresh. And it is an easy place to live . . . I mean there is nothing much to worry about. For instance, there is no rush for anything. But it is also a very boring place, so much that I did not even want to get out of bed because there was nothing for me to do.[6]

Some fared far worse than Lily, as in the case of Goldie's teacher husband, who traded his pens for a long pair of chopsticks and earned his living by the frying pan.

My study in 1991 found no 'Gucci Chinese' or 'yacht people', living in Canada as if they were in 'a little Hong Kong' (Canon 1989). Scarborough and its Chinatown became substitutes for the home they had lost; others lived in the suburbs because they had family there. For immigrant women who do not speak English, as in the case of one of my interviewees, the Chinese community indeed 'makes life bearable'. Removed from Hong Kong's familiar and non-intimidating cultural habitat, the immigrants faced a massive task of reorienting themselves to a new urban habitat, both the physical and the cultural dimensions of which were strange. As one college graduate said, 'I felt like a mentally handicapped person in the first year, and had to relearn everything from taking a bus to cooking with a stove'. Even for those who were well-educated, moving to Canada meant revising their expectations of life and re-ordering priorities. Place as bounded localities, place as activity spaces or nodes in activity networks, and the subject's positioning: all needed renegotiating and re-establishing, involving some initial disorientation, much determination, and flexibility in reassessing personal goals, not always without some regrets.

Returning to Hong Kong

One would think that those returning to their native homeland should be able to regain the sense of home. And that was what I thought when I returned in 1992 to Hong Kong, where I was born and spent most of my adolescence, after living in Toronto for ten years. My enthusiasm did not last long, however. I found that I could no longer engage in meaningful conversation with the few old friends who were still living in Hong Kong, and new friends did not share my world-view on many things. Without my being aware of it, a decade of living in Canada had obviously changed how I thought and felt. The feeling of not being 'at home' in my supposed 'homeland' prompted me to undertake another in-depth interview study, this time of eighteen Canadian returnees in Hong Kong, five men and thirteen women, between 1994 and 1996 (Chan 1996).

In the context of immigrant study, 'home' usually refers to the native land of a person or a person's ancestors. But few would dispute that 'home' is more 'an ideological construct' than simply a 'dwelling house' and is often 'associated with pleasant memories, intimate situation, a place of warmth and protective security' (Somerville 1992: 5).[7] Thus, the word 'home' always carries a positive connotation, as in 'homecoming', 'home-made food', 'home truth', etc. (Robertson *et al.* 1994: 94). We shall see in the later part of my discussion, however, that this positivist thinking is challenged by the post-modernist point of view.

Following the argument of some scholars that cultural identification is a contributing factor to one's feeling of 'at-homeness', I explored the existence of a Hong Kong culture and my subjects' identification with it. Culture here refers to a set of values 'as to the right way to live and work, what to teach [one's] children, and what [one's] public and private responsibilities should be' (Bellah *et al.* 1985: vii). First coined by de Tocqueville in describing the mores of a community and further explored by Bellah and his colleagues, the term 'habits of the heart' is thought to be one way to understand the culture of a society. This is a narrower, more personal and mundane way of defining culture than Said's: 'the . . . texts, traditions, continuities that make up the very web of a culture' (Said 1983: 6), but it is a practical one for my purposes here.

There was no doubt in the minds of my interviewees that there is a 'Hong Kong culture'. However, in one way or another, these Chinese-Canadian returnees led me to think that they were not part of it. They decided that they were different from and, more importantly, better than the '*Xianggangren*' (people of Hong Kong). Their criticisms dealt with materialistic goals, status-consciousness, ambition, avoidance of responsibility, lack of openness and of courtesy.

Xianggangren, they claimed, are too materialistic and too busy making money. Adam said, 'You can be a bum in Canada and still have a lot of friends, but not here. They measure you by the money you have and the brand name of the clothes you wear'. *Xianggangren* were thought of as being too rank- and status-conscious, too competitive and too goal-oriented. '*Xianggangren* are *jigongjinli*' (fixated on fast results and benefits), said Alex. 'My boss will phone me as soon as I am not reporting any results and will ask me sarcastically whether I am enjoying life so much that I forget to do my job', he added. For Sally, the shock came when she was called in by the school to take note that her daughter was below the class average. 'I was so surprised because my daughter already got over 80 per cent in her school work!' A few of my interviewees also thought of *Xianggangren* as shunning responsibility. For instance, Gary thought that *Xianggangren* were very '*sugu*' (figurative speech meaning they disappear when work is heavy). Echoing the same observation was Yvette, who, like Gary, thought that part of the reason for her heavy workload came from the fact that *Xianggangren* tend to avoid making decisions.

Worse still, *Xianggangren* are also perceived as unwilling to share information, selfish, impolite, rude and having no sense of fair-play. 'Can you imagine that it took two full days for a woman friend to tell me what her father did for a living? This would never happen with my overseas Chinese or foreign friends', said Adam. 'I had better friends in Canada than in Hong Kong because people here don't open up as much as they do in Canada', added Richard. As regards the lack of fair-play among *Xianggangren*, Alex gave the following vivid example:

> Just now, I was the first on the street waiting for a taxi; then a young guy appeared and stood in front of me; then a family of four stood in front of him. And when the family got on to a taxi, the young guy protested. I found myself wondering whether I should join in the protest. I did not. What is the use? This is Hong Kong!
>
> (Alex)

Clearly, many of my interviewees feel they are different from and, more importantly, 'better' than the average *Xianggangren*. In their opinion, they are more civic-minded, more honest, and, above all, they feel they espouse a more serious attitude towards work and fair-play. How has this rejection of aspects of their home culture come about? First of all, quite a number of my interviewees – as we shall see later in the discussion of ethnic identity – no longer identify themselves as *Xianggangren*, largely because of their migratory experience. Many of the above remarks, in fact, implicitly or explicitly made comparison to similar situations in Canada. On the other hand, there is no dispute that my interviewees' high education levels and higher professional positions must have made them feel a world apart from the majority of the *Xianggangren*, who earn probably only a fraction of what they earn and who live a very different life-style.[8] Thus, it is not surprising that Diana does not feel comfortable among crowds in the market and thinks that people cheat, and that Richard prefers living in Canada where they own a car, shop in spacious supermarkets, live in a large house, and where most of the Chinese they come across belong to their class.

Their sense of superiority might also be due to their success in migrating in the first place. As Hardie (1994: 64) pointed out, the successful emigrants felt that 'their worth has been demonstrated by an impassive authority and that they therefore qualify as part of the Hong Kong's élite'. The sense of superiority might be further heightened by their ability 'to make it out' in Canada, as evidenced in the following statement:

> I gained the exposure that I would never have if I did not leave, e.g. the Hi-Tech computing and communication equipment that I was able to work with . . . I would not have the confidence to take up my present job if I had not lived and worked in Canada.
>
> (Yvette)

I am inclined to think that my interviewees' sense of superiority, or even self-righteousness, may have something to do with the high value that Canadian society places on the worth of the individual. I remember vividly how my 9-year-old daughter changed in a short period of one year from a diffident Hong Kong girl to a confident Canadian. Any parents in Canada could describe how teachers tend to be encouraging, praising their pupils' efforts in school reports rather than focusing on grades.

To summarise, the returnees were very critical of Hong Kong and its people once they had returned 'home'. However, some mentioned their sense that Hong Kong had improved quite significantly in terms of public civility and living condition since the 1970s. 'I could not believe it when my husband told me that people in Hong Kong were lining up in the subway', said Sally, who originally had reservations about coming back to live because of her previous negative experience of living here. And many of them also hastened to tell me that all their complaints, including those about Hong Kong's overcrowding, pollution and dirt, though important, were 'minor irritants only' in their consideration of Hong Kong as home. As Yvette and Winnie explained:

> I think one's attitude matters a lot. When I was in Vancouver, I cannot say that I felt completely at home, but I did try to understand, as I do now. For instance, I constantly remind myself that there are more people here for less resources, so it is natural for people to be more aggressive.
>
> (Yvette)

> I used to tell my foreign friends that Hong Kong was a place of choice. Only after I emigrated to Canada, I became consciously aware of issues of social equity. Then I realised that for the majority of the people in Hong Kong, who barely make enough for their basic needs, there is not much choice.
>
> (Winnie)

In other words, their migratory experience had not only sharpened their perception of their home culture, but had deepened their understanding of it. The culture of a society is not static, of course, particularly that of a society as energetic and as sensitively located in geopolitical terms as Hong Kong. And, as Hong Kong's culture has changed in the years they were away, my interviewees admitted that they might have changed as well. 'Maybe I have changed too. For instance, it is without doubt that my sensitivity to things around me has sharpened because of my experience living in Canada', said Eva.

In sum, after being immersed in another culture, my interviewees could no longer identify with many of the values and world-views perceived to be part of the culture of Hong Kong. At the same time, being intelligent people, many of them were also trying to reach a level of understanding. Both Beth and Diana were embarrassed when I pointed out that their preference to live in Canada and their discomfort among the crowds of Hong Kong may be due to class sentiments. They replied that they would certainly feel more 'at home' if Hong Kong were not such a stratified society. All in all, not being able to identify with Hong Kong culture adds to some of their other major concerns, such as political stability and social order, congested living condition, long working hours and the general quality of life, as they try to make Hong Kong their home once more.[9] And, in their struggle to find and make a home for themselves, the question of identity emerges as an obvious consideration.

Identity and home

Ceded to British rule one-and-a-half centuries ago, Hong Kong (just less than 1,000 square kilometres, population 6.6 million in 1997) has never been considered a secure separate entity from China, but a 'borrowed place' on 'borrowed time'.[10] It has been, for example, just as much an entrepôt for Chinese migrants as for Chinese goods. Until 1940 there was no restriction on the human flow between Hong Kong and the Chinese mainland. The communist take-over in mainland China in 1949, together with turbulent events on the mainland in the following decades, brought floods of Chinese refugees into the colony. It is no surprise, then, that as late as 1991, the number of foreign-born in the territory was as high as 2,222,684 out of a total population of 5,522,281 with the majority claiming ancestral roots on the mainland (Hong Kong 1991 Population Census, quoted in Skeldon 1995: 81). At the time of writing, while over 60,000 leave Hong Kong each year, the population is replenished by an equally large number of immigrants from the mainland (more than 150 per day) not to mention returned migrants and immigrant foreign nationals.

Thus, for the *Xianggangren*, Hongkongers, questions of identity are particularly complex. Lau and Kuan (1988) described them as 'refugees in their own homeland' and Siu (1996: 178) suggested that 'the cultural reference point is often the imagined native place of the deceased carved in stone, be it China, India, or Britain'. They had never been *huayi*, overseas Chinese, and as the territory approached its destiny to be part of the People's Republic of China in 1997, most preferred 'exit' over 'voice' (Wong 1992: 930) and pursued the survival strategy that Siu-kai Lau termed 'utilitarian familism' – 'culturally and economically conservative, with a strong dose of political apathy towards the larger society which they helped make' (Lau, 1985, quoted in Siu 1996: 182).

Siu (1996) argued however that there is a distinctive Hong Kong identity, particularly among the post-war baby-boomers. To these baby-boomers who were born after the Chinese communist take-over in 1949, who were educated in an education system modelled on that of Britain, and who had no contact with their motherland until the 1970s, Hong Kong is home. They have been British colonial subjects without democratic rights but with all other personal freedoms – including travel – and an open and just legal system. Their identity as *Xianggangren* has been sharpened as interaction with the Chinese mainland in the last two decades makes them realise how different they are from their fellow Chinese across the border. The situation is further complicated by the large number of immigrants from the Chinese mainland arriving in the territory every day (see figures above). Many now 'draw boundaries between "we" the Hong Kong *yan* [*Xianggangren*], and "they" the "new immigrants"' (Siu 1996: 187; Nadel and Woo 1996). Hence, many have taken steps to ensure they can settle elsewhere if life in Hong Kong becomes unacceptable to them now that it has been once more incorporated into the political entity of the communist Chinese Republic. As a result, about one-sixth of the population (i.e. about one million people) has the right of abode in another country (Rosario 1995). Many of these foreign passport holders have been educated overseas – they have been brought up largely in a culture that is neither that of the Chinese mainland, nor of Britain, nor of any other country whose passport they may hold – Canada, Australia, New Zealand or the USA, for example. Hong Kong, this vibrant and dynamic place has, as a nation, never existed. Today, it is a mere Special Administrative Region of the People's Republic of China. How a Hong Kong Chinese chooses to identify him- or herself in terms of nationality is, thus, a highly sensitive issue for both citizen and state.

It is unexpected, therefore, that most of my interviewees seem to be quite sure of who they are *vis-à-vis* ethnic identity, which they define, in one way or the other, in terms of the influence of Chinese culture in their life or of their concern with Hong Kong and China. For instance, Alex has decided that he is a *Zhongguoren* (Chinese) because of the values he holds. He explained to me that he strongly believes in the institution of the family and parental responsibilities, which he thinks is a result of the example set by his father. Alex worked during his teens at a restaurant until 3–4 a.m. His father would wait for him until the early hours in order to walk him home. Now, Alex takes Chinese philosophy as his guide in life and reads *The Art of War* (a Chinese classic on the tactics of war) for inspiration on how to do business. Another respondent, Richard, explained his Chineseness to me in terms of *gen* (roots). He feels that knowledge about our *gen* is very important. Although he studied in an English school, he likes to read Chinese literature, history and poetry. Richard, like Alex, was influenced by his father, who worked for the Hong Kong government, was very conscious of Hong Kong's colonial status and made sure that Richard was aware of it as well. To both Alex and Richard, being *Zhongguoren* means possession of some perceived Chinese peculiarities and caring sentiments towards China. In contrast, Adam has decided that he is 'a Canadian' and 'a foreigner' in Hong Kong because, in his own words, he 'thinks differently and does things differently'.

Yet in affirming their ethnic identity, many of my interviewees also emphasised the separation of ethnic identity from nationality. Alex defined himself as '*Zhongguoren* with a Canadian passport', but emphasised that he is 'not thinking along nationalistic lines'. Yvette thought of herself as 'a *huaren*' (person of the Chinese race) with a 'Canadian nationality'. Even Eva, who strongly felt that she is *Zhongguoren* and who came back as soon as she finished university, qualified her Chinese identity by adding that it does not mean that she identifies with the government on the Chinese mainland or any Chinese government. Indeed, a few of my interviewees even refused to identify themselves as *Zhongguoren* because they did not want to have anything to do with Chinese nationality.

For these respondents, the migratory experience has brought about a self-examination of values and of cultural roots. Ethnicity, culture, geography, politics and nationality are actively in question during this process of introspection, as they mull over their roots, trace out the routes they have taken to their present sense of personal and cultural identity (Hall 1990). Layer upon layer of socialisation, choices made, and experience, underlie their self-concepts. Their sentiments, however, are not unique; they reflect the attitude of many overseas Chinese. As Tu (1991: 1) observes, the current phase of emigration of highly educated ethnic Chinese for educational or political reasons has given Chinese ethnicity a new meaning. Being Chinese now is 'intertwined with China as a political concept and Chinese culture as a living reality – the cultural China'. The use of '*huaren*' by overseas Chinese from the 1980s instead of *Zhongguoren* comes from this idea, as can be seen in the following definition of *huaren*:

> [*Huaren*] designates people of a variety of nationalities who are ethnically and culturally Chinese. *Huaren* is not geopolitically centred, for it indicates a common ancestry and a shared cultural background, while *Zhongguoren* necessarily evokes obligations and loyalties of political affiliation and the myth of the Middle Kingdom.
>
> (Tu 1991: 22)

In fact, a few of my interviewees told me that, if possible, they would like to be identified as *diqiuren* (people of the earth). However, since I did not allow them to do so, they decided that their sense of identity should be made culturally, i.e. through what touches their heart and influences their thinking and behaviour as discussed above (Bellah *et al.* 1985). Therefore, except for Richard, who expressed uncomfortable feelings about his decision to emigrate *vis-à-vis* his feeling of being *Zhongguoren*, most of my interviewees did not express any misgivings about their original decision to leave Hong Kong insofar as their ethnic identity is concerned. They do not think that ethnicity should be tied to nationality, nor to obligations and rights of citizenship. Eva elaborated on this:

> Our obligations and duties should be placed where we live and we should act on what we decide to be morally right or wrong . . . Therefore, I am for the separation of Quebec as much as the independence of Tibet, if that is the wish of the people living in both places.

The relatively favourable climate for minority rights and preservation of ethnic culture in Canada has also created the 'hyphenated Chinese', as many of my interviewees would like themselves to be called, i.e. Chinese-Canadian. They migrated primarily because of political concerns, but also because they would like to choose where to live. Canada may not be the ideal place, but it is good enough, particularly because Canada separates citizenship and residency, and allows them to return to Hong Kong to live. However, it should be noted that the above interviews were carried out before the announcement of a decision by the Chinese government that returnees in Hong Kong who desire consular protection must declare their foreign citizenship (*South China Morning Post*, 25 March 1996). In other words, if my interviewees opt for consular protection from Canada, they will have the right to live and work in Hong Kong, but not the 'right of abode' – the minimal definition of home – as they have in Canada. They will be treated as other foreign workers in Hong Kong, and will 'be expelled if they fall foul of the law' (*South China Morning Post* 18 April 1996). By choosing to live in Hong Kong, Chinese-Canadians now find that 'home' in Hong Kong offers only a qualified sense of security and legitimacy.

Where is home? Everywhere or nowhere?

If both Hong Kong and Canada are not quite home, where is home? The answers to this question offered by my interviewees are full of contingencies:

> Home will be where I can pursue my life priorities at any point of time: a place where I have a purpose to be there, but that place is not necessarily home in the idealistic sense, if you understand what I mean.
>
> (Ted)

> For myself I will make Hong Kong, or anywhere else, home as long as there are opportunities.
>
> (Alex)

> It will all depend on how things work out here, or anywhere else. The most important aspect of home to me will be a personal space where I can be my private self and freedom to do what I like.
>
> (Adam)

> Although I'd like to make home in Paris or New York because of my interest, I know that being Chinese and an amateur dancer, I would not be able to make a living there. So I will be here as long as I have a job that affords me to indulge in my interests in dancing and fashion.
>
> (Chris)

There is a sharply drawn gendered difference in responses, especially in the case of married women. For women who are mothers and wives, home is invariably where one's spouse or children are:

> As a woman, at least as far as I am concerned, home is where my closest family is, i.e. where my children and husband are. Or I should say, where my husband is because my children will grow up and leave.
>
> (Lily)

> As much as home is concerned, it will always be where my husband and daughters are. Hence, if my daughter needs me in Canada or anywhere else, I will go to her too.
>
> (Yvette)

> If you asked me where is home, I would say that it is where my husband and children are. Therefore, I may say I have two homes: one in Canada where my children are and one in Hong Kong where my husband is.
>
> (Diana)

It is clear from the above responses that the answer to the question of 'where is home' is dictated by contingencies of one kind or another, personal as well as circumstantial. And the migratory experience has led many of my interviewees to be more critical of the assuredness of home. For instance, as soon as I informed Goldie of the title of my study, she told me that she 'had also been thinking a lot about the guarantee of home lately'. Diana had been too.

> After migration, I truly feel that real security is in God; only He provides the most dependable security, which cannot be had in this world . . . I am not alone. Many immigrants turned to God because they lost their roots and began to search for the spiritual roots . . . I can now be at peace wherever I am and whatever I encounter.
>
> (Goldie)

> Sure, there were handicaps and losses particular to migrating to and living in a foreign country, but I think my Christian faith helps a lot in this regard. For the Bible told us that this world is not our home – a feeling that I have been acutely aware of in these two years when I had to live between Hong Kong and Toronto.
>
> (Diana)

It seems that Goldie, Diana and others like them have given up the search for home, at least for the idealised, earthly home. The traditional definitions of home have failed them and many of them began to sound like post-modernist travellers talking about home nowhere and home anywhere. The post-modernist discourse on home challenges the traditional notion of home, and permits one to make some of the most explicit connections between migration and home because the world described by post-modernist thinkers seems to exemplify some of the important elements of the migrant's condition: a world full of unpredictable contingencies, and in constant flux. For example, Said's (1983) 'travelling theory' stands in direct opposition to the ideological construct of home. Whereas the concept of travelling theory accepts change as a natural state of affairs, home beseeches closure. Contrary to the breaking down of boundaries which is natural with the movement of travel, home in order to maintain its state of familiarity is necessarily a 'walled city'. Change foretells uncertainties and most often fear, while home gives guarantees and assurance. Which is better? For the homeseeker, surely the latter. For Said (1983: 230), the former because, in his view, it represents 'an opportunity for remaining skeptical and critical, succumbing neither to dogmatism nor to sulky gloom'. Trinh (1991) holds a similar view.

Some of my interviewees had not given up the search for home. As Ted wrote to me in the little notebook I gave him:

> You want to talk about home. Thanks. I feel so much affection for this word, sentimental, perhaps, that I nearly treated it as reality . . . I suppose I am an exile seeking home, '*tianyarubilin*' [the distanced sky is like my neighbour]. I think I'm a pretty bad guerrilla fighter – one without weapons.

Ted is yearning for an idealised past. He also is longing for a future inclusive of such a home.

> Well, we may be at a place to achieve some means and goals, or because of some bonds, but that place is not necessarily home. Tao Yuanming (a poet in ancient China) once told a story about some people who got lost and tried to rebuild a community with a single aim: to survive. Hence, it was a community without frills and politics.

Ted is however not alone in his pursuit of the idealised home. With others in my respondent group, he is searching for a home that includes 'shelter, hearth, heart, privacy, root, abode and (possibly) paradise' (see note 7). Emigration to Canada represented, to an extent, a journey in search of the perfect home. The circumstances necessitating their emigration and return, and their experiences since, have contributed to their growing sense that such a home is nowhere to be found in this world. In this sense, they are 'homeless exiles' in the post-modern world. But could they be 'home' anywhere, even according to the post-modernist view? Some thought they could:

> I don't think that we should seek belongingness because it will mean limited and missing opportunities. Accept differences and be comfortable with ourselves being who we are are more important than the search of home. If I may, I would like to make the world as my home.
>
> (Alex)

> After emigration, I found that home can be possible only when we have an ideal. Because only people with ideals will continue to ask, to think, and, in my case, continue to write . . . My recent awakening in this regard is: there is home if there are poems; there is love if there is home; thus, everywhere can be my home.
>
> (Goldie writing in the little notebook I gave her)

> I don't think we should be so hung-up on the idea of home. I will try to immerse in where I am, and try to find myself, then I am sure anywhere can be home.
>
> (Paula)

Are Alex, Goldie and Paula siding with Said (1983) and Trinh (1991) in seeing the idea of home as a restraining force, one that prevents them from accepting differences and things that are new? Are they, therefore, prepared to make home *anywhere* and *everywhere*? According to the definition of Agnes Heller (1995), my interviewees might be said to be 'geographically promiscuous' because they 'carry their home on their backs' and are prepared to be home *anywhere*. This possibility seems to be verified by the places that some of my interviewees frequented both in Hong Kong and in Canada:

> I don't think the places I go are typical hang-outs for locals. That is the best thing I like Hong Kong: the chance to meet people from other places, most of them are Caucasian and overseas Chinese. Go to the West World and you will understand what I mean. I take it as a shelter/home island against the realities of Hong Kong.
>
> (Adam)

> Vancouver has a large Chinese population so I can get almost anything done by Chinese, including building a whole house. As far as shopping

> and eating are concerned, certainly you would not feel that you are in a different culture in Richmond where I shop and eat. There you can even find Hong Kong-style café.
>
> (Lily)

> Few people could tell that I was educated and lived in Canada for more than ten years. I guess because I spoke Cantonese all those years at home as my parents can't speak English and I was quite in touch with what was going on in Hong Kong because of the Chinese magazine and video tapes that were easily available to me. I stayed home a lot and went out with my mother and her friends most of the time, mostly to places frequented by Chinese. I even played mahjong with them.
>
> (Paula)

Is the world developing into a 'global village', in the sense that urban societies such as Hong Kong, Vancouver and Toronto are capable of catering to different tastes and life-styles enabling recent Hong Kong immigrants to act like 'modern sojourners' (Skeldon 1994) who learn the codes for membership wherever they may be so that they can be at home *anywhere* and *everywhere*? I judge that many of the returnees in my study are not modern sojourners, so defined. Instead they are closer to being 'homeless exiles', i.e. home *nowhere* in the post-modernist sense except, of course, when they locate 'islands of home' where they can forget their state of homelessness.

Conclusion

> Migration is a one-way trip. There is no home to go back to.
>
> (Stuart Hall, quoted in Chambers 1994: 9)

The stories of my interviewees seem to support the post-modernist argument that, whether home is desirable or not, homecoming for the migrants is impossible; for migrancy 'involves a movement in which neither the points of departure nor those of arrival are immutable or certain' (Chambers 1994: 5). We are now living in a world of 'cross-cultural traffic' and, as the process of globalisation accelerates, we are 'increasingly confronted with an extensive cultural and historical diversity that proves impermeable to the explanations we habitually employ' (Chambers 1994: 2–3). Migrants must therefore constantly negotiate between an inherited past and a heterogeneous present. They live in 'a state of "in-betweenness" belonging neither in one place nor the other . . . [i.e.] in an indeterminate state of hybridity' (Rutherford 1990: 25).

However, my interviewees do not *celebrate* the state of 'homelessness', or identify with the post-modernist point of view that home is constraining, crippling, discriminating and exclusive. I believe nonetheless that they would support the post-modernist argument for a new meaning of home: a home that recognises differences. As Lary told me, he will stay in Hong Kong if the government is willing to listen to different voices. For only then, he said, will conflicts have a better chance of being resolved. Will this sort of attitude evolve into a new ideological construct of home? For I believe that every human being needs a home – if not in the sense of the old idealised version, then at least one based on acceptance of contrasts instead of sameness, strangeness instead of familiarity, with no hatred or fear as a result. This kind of home is not only important for the massive number of today's migrants, but will be needed by everyone in today's fast-changing world. For the present, however, there is no end in sight for the search for home.

Notes

1 Migration is moving from one place to another; immigration refers to entering a country as a permanent resident; and emigration is leaving one's own country to take up permanent residency in another (Turner 1987). Russell King (1995) provides a good overview of human migration through the centuries.

2 A total of 38,841 and 36,511 Hong Kong emigrants arrived in Canada in 1992 and 1993, respectively (Lary 1995: 5).

3 Wickberg 1983; Canon 1989; Huang and Lawrence 1992; Man 1997. For studies of attitudes of Hong Kong Chinese to migration before the 1997 handback from Britain to China, see Li *et al*. (1995) and Findlay and Li (1997).

4 The number of migrants returned to Hong Kong from Commonwealth countries such as Canada, UK, and Australia cannot be ascertained accurately because of documentation procedures prior to 1 July 1997. Government estimates indicated that well over 12 per cent of those who emigrated might have returned, but some reports placed the figure of returnees as high as 16 per cent (South China Morning Post 15 February and 13 July 1994). That is to say, 23,000 to 31,000 of the 196,851 people who emigrated to Canada in the 1980–92 period might have returned to the territory. At the time of my study in 1996, only two studies were found to be remotely related to return migration in Hong Kong. Swanson (1995) looked at the possibility of return migration among the emigrants, while an interview project by Li et al. (1995) touched on the question of return migration. Neither dealt with the life experience of the returnees.

5 Due to the fact that it was an undergraduate project, with constraints of both time and resources, my sample was not meant to be fully representative of the Hong Kong immigrant population at the time. In-depth interviewing was appropriate, as the objective of the study was to gather data on the subjective experience of migration. For interviews as a research methodology, see Hammersley and Atkinson (1993: 112–126).

6 Pseudonyms are used throughout and fieldnotes were written in English shortly after the interview. Hence, the quotations represent my interpretation of the substance of ideas expressed by my interviewees rather than verbatim speech.

7 Home as shelter connotes the material form of home, in terms of physical structure which affords protection to oneself . . . Home as hearth connotes the warmth and cosiness which home provides to the body . . . Home as heart is very similar, but in this case the emphasis is on emotional rather than physiological security and health . . . Home as privacy involves the power to control one's own boundaries . . . Home as roots means one's source of identity and meaningfulness . . . Home as abode corresponds to . . . the minimal definition of home, that is, anywhere that one happens to stay, whether it be a palace or a park bench . . . Finally, home as paradise is an idealisation of all the positive features of home fused together (Somerville 1992: 530–533).

8 Hong Kong is a very stratified society with 20 per cent in the lowest income brackets making an average of HK$5,500 per month only, while the richest 20 per cent earn ten times more at an average of $49,250 per month in 1996 (*South China Morning Post* 6 December 1997). For this reason, Hong Kong has two living cost indices, one for the poor and one for the rich.

9 Although Hong Kong ranked sixth in its per capita gross domestic product at HK$189,985 in 1996, surpassing Britain, Canada and Australia, it ranked a mere twenty-second in a United Nations index for quality of life versus Canada (second) and Australia (seventh). For example, Hong Kong people work an average of 9.7 hours per day to rank third after South Korea and Chile and, because of the exorbitant price of land, many cannot afford to buy a tiny apartment which they can call home. These conditions underlay the fact that none of my interviewees ruled out the possibility of resuming their residency in Canada.

10 The phrase was coined by Richard Hughes (1976) in his book with the same title. For a short history of Hong Kong, read Jan Morris (1989) and Frank Welsh (1997).

References

Bellah, Robert N., Madsen, R., Sullivan, W.M., Swidler, A. and Tipton S.M. (1985) *Habits of the Heart: Individualism and Commitment in American Life*, Berkeley: University of California Press.

Canon, Margaret (1989) *China Tide: The Revealing Story of the Hong Kong Exodus to Canada*, Toronto: Harper and Collins.

Chambers, Iain (1994) *Migrancy, Culture, Identity*, London and New York: Routledge.

Chan, Wendy (1996) *Home But Not Home: A Case Study of Some Canadian Returnees in Hong Kong*, unpublished M.Phil. thesis, Hong Kong University of Science and Technology.

Chan Tang, Wendy (1991) *Lifestyle Change as a Result of Immigration: Recent Immigrant Women from Hong Kong*, unpublished undergraduate thesis, University of Toronto.

Findlay, A.M. and Li, F.N.L. (1997) 'An auto-biographical approach to understanding migration: the case of Hong Kong emigrants', Area 29, 1: 3–44.

Hall, Stuart (1990) 'Cultural identity and diaspora,' in Jonathan Rutherford (ed.) *Identity, Community, Culture, Difference*, London: Lawrence, 222–237.

Hammersley, Martyn and Atkinson, Paul (1993) *Ethnography: Principles in Practice*, London and New York: Routledge.

Hardie, Edward T.L. (1994) 'Recruitment and release: migration advisers and the creation of exiles', in Ronald Skeldon (ed.) *Reluctant Exiles? Migration from Hong Kong and the New Overseas Chinese*, Hong Kong: Hong Kong University Press, 52–67.

Heller, Agnes (1995) 'Where are we at home,' *Thesis Eleven* 41: 1–18.
Huang, Evelyn and Lawrence, Jeffrey (1992) *Chinese Canadian: Voices from a Community*, Vancouver: Douglas and McIntyre.
Hughes, Richard (1976) *Borrowed Place, Borrowed Time: Hong Kong and its Many Faces*, London: Andre Deutsch.
King, Russell (1995) 'Migrations, globalization and place', in Doreen Massey and Pat Jess (eds) *A Place in the World?*, Oxford: Open University and Oxford University Press, 6–33.
Lary, Diana (1995) '1993 Hong Kong immigrants landed in Canada: demographics', *Canada and Hong Kong Update* Winter: 5–6.
Lau, S.K. (1985) *Society and Politics of Hong Kong*, Hong Kong: Chinese University Press.
Lau, S.K. and Kuan, H.C. (1988) *The Ethos of the Hong Kong Chinese*, Hong Kong: Chinese University of Hong Kong Press.
Li, F.N.L, Jowett, A.J., Findlay, A.M. and Skeldon, R. (1995) 'Discourse on migration and ethnic identity: interviews with professionals in Hong Kong', *Transactions of the Institute of British Geographers* 20, 3: 342–356.
Man, Guida (1997) 'Women's work is never done: social organization of work and the experience of women in middle-class Hong Kong Chinese immigrant families in Canada', *Advances in Gender Research* 2: 183–226.
Michelson, William and Reed, Paul (1975) 'The time budget', in William Michelson (ed.) *Behavioral Research Methods in Environmental Design*, Stroudsberg, Pennsylvania: Dowden, Hutchinson and Ross, 180–234.
Morris, Jan (1989) *Hong Kong*, New York: Vintage Books.
Nadel, Alison and Woo, Anthony (1996) 'Fear and loathing – those are the emotions some Hong Kongers feel for mainland immigrants', *Far Eastern Economic Review* 31 October: 44–46.
Robertson, George, Mash, Melinda, Tickner, Lisa, Bird, Jon, Curtis, Barry and Putnam, Tim (eds) (1994) *Travellers' Tales: Narratives of Home and Displacement*, New York: Routledge.
Rosario, Louise do (1995) 'Hong Kong: futures and options: up to a million local Chinese could opt to leave', *Far Eastern Economic Review* 15 June: 23.
Rutherford, Jonathan (1990) 'A place called home', in Jonathan Rutherford (ed.) *Identity, Community, Culture, Difference*, London: Lawrence.
Said, Edward W. (1983) *The World, the Text, and the Critic*, Cambridge, Massachusetts: Harvard University Press.
Siu, Helen (1996) 'Remade in Hong Kong: weaving into the Chinese cultural tapestry', in Tao Tao Liu and David Faure (eds) *Unity and Diversity: Local Cultures and Identities in China*, Hong Kong: Hong Kong University Press, 177–196.
Skeldon, Ronald (ed.) (1994) *Reluctant Exiles? Migration from Hong Kong and the New Overseas Chinese*, Hong Kong: Hong Kong University Press.
—— (1995) *Emigration from Hong Kong*, Hong Kong: Chinese University of Hong Kong.
Somerville, Peter (1992) 'Homelessness and the meaning of home: rooflessness or rootlessness?', *International Journal of Urban and Regional Research* 16, 4: 529–539.
South China Morning Post, 15 February, 13 July 1994; 25 March and 18 April 1996; 6 December 1997, Hong Kong.
Swanson, L. (1995) Hong Kong migrants to Canada – a micro-analytical approach, M.Phil. thesis, University of Hong Kong.

Trinh, Minh-ha (1991) *When the Moon Waxes Red*, London and New York: Routledge.

Turner, G.W. (1987) (ed.) *Australian Concise Oxford Dictionary*, Melbourne: Oxford University Press.

Tu, Weiming (1991) 'Cultural China: the periphery as the centre', *Daedalus*, 120, 2: 1–32.

Wang, Gungwu (1991) *China and the Chinese Overseas*, Singapore: Times Academic Press.

Welsh, Frank (1997) *A History of Hong Kong*, London: Harper Collins Publishers.

Wickberg, Edgar (ed.) (1983) *From China to Canada*, Toronto: McClelland and Stewart.

Wong, Siu-lun (1992) 'Emigration and stability in Hong Kong', *Asian Survey* 32, 10: 918–933.

12

EMBODYING OLD AGE

Richard Hugman (Australia)

> It is as though, walking down Shaftsbury Avenue as a young man, I was suddenly kidnapped, rushed into a theatre and made to don the grey hair, the wrinkles and other attributes of age, then wheeled on stage. Behind the appearance of age I am the same person, with the same thoughts, as when I was younger.
>
> (Novelist and playwright J.B. Priestley, at 79 years of age, quoted in Featherstone and Hepworth 1989: 148)

Old age is a social and cultural construction. Such words as 'aged', 'elder', 'elderly' or 'old' are used unproblematically in everyday language. These terms are applied both to people and to objects, in ways that convey a deep sense that the passage of time is a key feature in establishing identity (Hazan 1994). In this sense 'old age' is often equated with personal chronology (how old a person is in years). In some circumstances old age may be imbued with positive values. Providing 'links with the past', a sense of 'continuity' or 'history', representing 'tradition', 'heritage' or being 'venerable', are all facets of old age as good. At the same time, old age is also closely associated with decline, decay, decrepitude and death.

Given the contested and contradictory meaning of old age, much writing in the field of social gerontology has been devoted to charting and understanding the processes of ageing and being 'old' (Featherstone and Hepworth 1989; Bond *et al.* 1993; Hugman 1994; Bytheway 1995). In particular, more recent discussion has focused on the wide variation in ageing and in being 'old' experienced by people according to diverse social factors such as gender, race and cultural background, disability, sexuality, social class and so on (Biggs 1993; Blakemore and Boneham 1994), as well as physical health (Anderson 1992; McCallum 1997).

Yet despite the ambiguous, even contradictory, nature of old age, and the wide variation in experience of later life, policy makers and practitioners in health and social welfare are ever more closely focused on older members of society in the development and provision of services (McCallum and Geiselhart 1996). As the demographic profiles of industrial societies are rapidly ageing (Grundy and Harrop 1992), there is a clear focus on old age as the crucial dimension in the development of health and welfare policy and practice (Sax 1993; Ozanne 1997). Indeed, it has been argued that attention to these questions must be the central basis of planning for social and political stability in western societies in the next century (Esping-Andersen 1990; Thurow 1996).

While recognising that the implications of shifting demographics for health and welfare infrastructures and practices are both substantial and far-reaching, it must also be noted that it is only a *minority* of older people who actually use such services (Anderson 1992; Hugman 1994; McCallum and Geiselhart 1996). At any particular point in time, it will be less than one-third of the population aged over 65 years who have a sufficient degree of frailty or ill-health so as to require either assistance or support from others in their daily lives or the intervention of professional services (Hugman 1994: 125). Although ageing is a major factor in the incidence of disability and ill-health, the overall rates are less than 50 per cent below the age of 85 years and the incidence does not begin to increase rapidly until the age of 75 (Office of Population Censuses and Surveys 1988; McCallum 1997). In other words, until reaching the age of 85, there is a greater than even probability that a person will not experience disability or ill-health. However, as McCallum (1997: 66) points out, incidence and impact are not the same. For example, organic conditions such as forms of dementia (including Alzheimer's disease) occur relatively infrequently, but have a very profound impact on the life of the individual and the immediate community of family, friends and neighbours.

Im-age

Observations such as that above concerning the impact of disease perhaps may provide one clue in attempting to explain the durability of negative stereotypes of older people. The sheer impact of certain age-related events and conditions may give them particular power within the formation and perpetuation of images. That it is approximately one person in every twenty who develops dementia is obscured by the quantity of resources and effort that have to be applied to the provision of an appropriate response. The other nineteen older people thus are rendered invisible and the strong association of old age with decrepitude is reinforced.

It is also the case that *normal* human ageing does involve change in the physical, psychological, social and spiritual aspects of each individual person (Bond *et al.* 1993; McCallum 1997). Physical changes in the person are those which are most immediately tangible to others. They are seen in features such as skin and hair, body shape and structure, and – less immediately apparent – changes in sight and hearing (Briggs 1993). The other area of physical ageing that has a profound implication for social images of old age, especially for women, is in relation to reproduction (Greer 1991). It is at this point that the interconnectedness of physical ageing with psychological, social and spiritual aspects must be made explicit. The process of ageing includes change in all aspects of life, and the loss of the capacity to bear children, while physical, has a meaning that is constructed socially, psychologically and spiritually. Other events and processes of ageing can be understood in the same way. Indeed, change in each of these aspects of personal existence and experience normally continues throughout one's life. It seems unlikely that J.B. Priestley, quoted at the start of this chapter, *actually* thought in exactly the same way at age 79 as he had at 29. However, it is important to recognise that the impact of physical change should not be taken out of context, which has cultural, geographical and historical dimensions. What was important for Priestley was that he did not own the aged identity constructed for him by his social circumstances. He had gone through a prolonged *rite of passage* (van Gennep 1960) that is simultaneously a *taboo* in modern Western culture: *he had grown old*.

Taken together, the physical effects of *normal* ageing have come to be characterised as the loss of personal attributes that are of importance in modern industrial society. Old people are seen to be 'past it' for both production and reproduction. As Phillipson (1982) has argued, modern old age therefore is primarily a capitalist phenomenon. That the number of older people who have personal financial resources (through superannuation, for example) continues to grow, and that this appears to be a means through which some social power may be exercised (most notably through consumption), does not necessarily challenge this interpretation. Such power is still based on economic status and thus is linked to social class. Extreme examples can be used to create stereotypes which are not representative merely of small minorities, but also provide the basis for the construction of dichotomies that become the foundations for discriminatory policies and practices. In this context, the dichotomy is between 'the crumbly' (to borrow a particularly oppressive idiom) on one side and 'the disco-granny' on the other, while the majority of older people cannot be located in either category.

Despite the 'normality' of ageing, and the diversity experienced within this process, the culture dominant in advanced industrial societies espouses youth as the 'norm'. This culture is constructed around images in which are embedded ideas of beauty, attractiveness, power and desirability founded on images of youth (Bytheway 1995). Women are treated differently in these terms, confronting a particular emphasis on what it is to 'look good' (Sontag 1978). Even in the era of 'power dressing' and the supposed levelling between the sexes, in how an older person may present as 'attractive' it may still be easier for older men to be taken seriously in the public arena in this regard, as sexism cross-cuts ageism (Woodward 1991; Greer 1991).

Several writers (Fennell *et al.* 1988; Featherstone and Hepworth 1989; Biggs 1993; Bytheway 1995) have argued that one powerful perpetuation of negative images of older people is found in the attention paid by policy makers, professional practitioners and academic researchers to deficit, disease, disability, decrepitude and decay. Through a concentration on problems, the twin effects of institutionalising ageism in this way are that older people are simultaneously lumped together as a 'social issue' while being individualised in terms of the responses to be made. 'Older people' are undifferentiated in the scope of theories and research, while practical attention is paid only to individual people on a case by case basis and rarely to the structural issues faced by older people collectively. As Biggs (1993: 37) points out, this leads to a situation in which it is not even the whole person but only the problematic aspect of the person's life to which a response will be addressed. Uniform responses in policy or practice based on chronological age alone will negate appropriate attention to difference and at the same time it may serve to obscure other social divisions that are discriminatory. Then, finally, at the social level the continuing membership of and positive contributions to the community by older people are denied by this process of 'pathologising' later life.

One facet of this process can be seen in the way that forms of health and social welfare provision are constructed within particular spaces (Willcocks *et al.* 1987; Biggs 1993; Hugman 1994; Laws 1997). The meaning of space here is both physical (location) and social (milieu) at the same time. Just as older people are marginalised and excluded socially in the ways that have been discussed above, so they may also be marginalised and excluded by institutionalised responses to their needs. The rest of this chapter will examine the ways in which *images* of old age are managed (created, sustained and changed) through the connection of physical and social space as the location for the management of older *people*. Three specific institutional forms of provision for older people ('residential homes', 'retirement communities' and 'home and community care') are taken as examples to explore the embodiment of old age as a social construction bounded in space and time.

Man-age

Residential care

Within the concept of 'residential care' are subsumed a range of institutional forms that vary, subtly between different parts of any one country, and more clearly between countries (Hugman 1994). Included are nursing homes and social care homes (variously called 'old people's homes', 'aged-care hostels', 'rest homes', 'retirement homes' and so on). The common criteria are that daily life takes place in a context that is shared with people with whom one might not have chosen to live, and that others (staff) exercise power over even the most personal aspects of daily routine. In the words of Willcocks *et al.* (1987), residential care means living 'private lives in public places'.

Despite continuing critique, and considerable effort, on the part of government agencies, academic researchers, professionals and service users (Clough 1981; Booth 1985; Willcocks *et al.* 1987; Newman 1988; Sinclair *et al.* 1990; Allen *et al.* 1992; Bartlett 1993; Peace *et al.* 1997), residential care for older people contains many of the factors identified with the 'total institution' (Goffman 1968). These factors include separation of a group literally and/or symbolically from the rest of society, limitation or denial of individuality, concentration of all aspects of life in one place, and control by staff of the use of time and space by residents. Identity is spoiled through depriving the individual person of the capacity to maintain a separation of 'front of house' (public) and 'backstage' (private) aspects of life (Goffman 1971). So, although official policies may emphasise 'domesticity', life in residential care for older people continues to have similarities with prison or the poor-house (Hugman 1994; Laws 1997 – cf. Foucault 1977; Scull 1979).

It is the relationship between space, time, action and identity which makes the conflation of residential care settings with the total institution so durable. This point may be illustrated by looking at three aspects of the way residential care impacts on the lived reality of older people. First, for those older people whose physical ageing is experienced as bodily decline, the use of space to manage identity can be important. Willcocks *et al.* (1987: 7) note that this process may take the form of concealment from others of the strategies adopted to accomplish daily living tasks, thus preserving an integrated sense of self as continuing to live 'as normal'. Personal routines can be modified to accommodate bodily change without undermining the contribution to identity of being able to perform as a competent member of the society. This is not to deny the psychological 'threat' or sense of loss experienced in making such adjustments (Biggs 1993), but to note that self-control over the links between action and identity can be gained from independence in the use of space and time. Limitations on this self-control, through the management of space and time by staff in residential care, challenge the identity of those older people who are residents, and so reinforce cultural associations between old age and dependence.

Second, residential care as managed spatial relations may affect the identity of older people as members of communities, comprising family, friends and neighbours. For those older people who need to enter residential care, community contact may be disrupted because of the symbolic (and sometimes quite tangible) barriers imposed through spatial factors in the organisation and provision of care. For example, if a residential care facility does not have sufficient space for private conversation, or sets limits on times for visiting, or does not enable the resident to go out to make visits to others, then family, friends and neighbours may experience difficulty in maintaining normal contact (Sinclair *et al.* 1990). Even when such social interaction is possible, it may still be that visitors have a sense of being 'in someone else's territory' that is a reflection of the resident being 'in a home', where they may feel 'at home' but not 'in *my own* home' (Willcocks *et al.* 1987). Family relationships and other friendships may thus become strained, if not fractured, further weakening the identity of the older person as a member of the community. So the apparent 'failure of the family' can, from this perspective, be seen as a consequence and not a cause (through continually rising demand) of residential care provision (Peace *et al.* 1997).[1]

Third, the process and organisation of residential care simultaneously groups all older people with care needs together – 'congregation' – and separates them from the rest of society – 'segregation' (Wolfensberger 1972). For older people this recreates and reinforces the common-sense notion (which for a time found parallels in social gerontological theory) that it is normative for older people to 'disengage', and hence withdraw, from society. More recently, this view has largely been replaced with a recognition that where older people do withdraw it is more likely to be the consequence of exclusionary social processes and structures rather than a 'natural' aspect of ageing (Hugman 1994: 21–23). Added to this, the concrete (sometimes quite literally) structures of the built environment communicate the interplay of congregation/segregation and make a profound statement that *people here are different*. Strategies such as the removal of institutional name boards cannot in themselves cover over the purpose of a building designed to house thirty or more people in a culture where 'homes' are normatively for nuclear families. Nursing 'homes', rest-'homes' and aged-care 'homes' cling to the notion of 'home' precisely because they are not 'home' in this sense.

In combination, these three aspects of residential care are central to understanding one very influential space/time/action/identity mix of old age. Old age in these places is constructed around dependence, both physically and socially. The tending of the elderly body in these institutions creates and sustains an identity of the elderly person as dependent, docile, accepting and (most preferred by staff) grateful (Biggs 1993: 154–155). The current author's experience, over more than a decade and in two countries (Australia and the UK), that social work and human service students routinely overestimate the proportions of older people living in residential care (often by a large degree) suggests anecdotally that such institutions define a normative view of older people. In so doing, these students appear to be operationalising professional perspectives on the identity of older people that are shared with other occupational groups (Latimer 1997).

is based on their being exclusive. These facilities are for the *affluent* older population and as such guarantee that niche markets are maintained. Second, because they emphasise the Third Age active embodiment of the older person, either overtly or by subtle management of contrary indicators, these facilities recreate a commodified social form of the 'neutral/male body' discussed by Harper (1997: 164–169). That is, they project an image of the older body as one where the person is in control and by implication there is no deterioration or decay. Where this does occur it can be managed through decor, design and an extensive social programme. As Harper points out, feminist theory of the body has demonstrated that this is an element of patriarchal society, in that by inference the female/older body is seen to be that which is not controlled by the person but has to be managed (medically or socially). Taken together, these two points suggest that the development of retirement communities encapsulates strong features of structural social divisions and so its post-modern cultural style easily enmeshes with patriarchal late-capitalism.

Home and community care

Of course, as has been noted above, the majority of older people in Western societies do not live in either residential care homes or retirement communities. For most older people, residence continues in the settings in which they already live, or in which anyone could live irrespective of their age (McCallum and Geiselhart 1996; Peace *et al.* 1997). Research evidence has continued to suggest that this is the preferred option, indeed that of choice wherever possible (Willcocks *et al.* 1987; Allen *et al.* 1992). For older people as much as people of any age, home is bound up with identity. It is the location in which, through the use of time and space, self-hood is enacted in the routines and decisions of daily living. Yet the common image of old age is not that of the home owner or tenant managing daily life as normal. Indeed, for much of the social gerontological literature, home in this sense has become another site for the enactment of old age as a part of the life-course defined by dependence.[2] This is the terrain of 'home and community care'.

Here, too, there are subtle differences of language: 'home and community care' is an Australian phrase (McCallum and Geiselhart 1996; Ozanne 1997). In New Zealand and in the UK 'community care' is the common term (Green 1993; Hugman 1996), contrasting with 'home care' in France (Henrard *et al.* 1991) and in Canada (Toner and Kutscher 1993). In the USA these terms are subsumed within the medicalisation of old age as part of 'health policy' so that 'in-home care' has different connotations (Newman 1988; Torres-Gil 1992). What is displayed in this use of language is that 'home' and 'community' are places in which the fabric of daily life is woven around 'care'. Care in this sense is the performance of tasks that tend to the needs of the person (Ungerson 1983). It may, or may not, encompass care as the expression of emotional warmth. However, the reassertion of ideas of home and/or community in policy discourse has been easy because it resonates with the 'common sense' of older people for whom these are the places of self (Allen *et al.* 1992). Home is not only where the heart is, but also where we can 'be ourselves' and 'do our own thing'.

The ambiguities encapsulated in the combination of 'home', 'community' and 'care' become more evident when it is shown that home and community care services, even if seen as 'social', 'domestic' and so on, are almost always provided for 'health' reasons (Toner and Kutscher 1993; McCallum and Geiselehart 1996). This is most clear when the tasks concern getting someone out of bed, toileting, bathing, dressing, but even the tasks of preparing meals, shopping or cleaning the house, will be carried out for a person who is physically or mentally 'frail' or 'infirm'. Social and emotional needs are unlikely to be met in this way. This meaning of home and community defines old age in terms of dependence as much as does residential care or other segregated living. Identity and self-hood at home and in the community are thus moulded in relation to the actions of others in performing care while also, perhaps more implicitly, intervening in the management of time and space.

Scull (1984) suggests that the dynamic of the move away from large-scale congregate forms of health and social welfare may be as much a consequence of the development of new 'technologies' for managing society as from an ideological rejection of institutional forms of care. The specifics of these technologies, seen as professional skills, organisational structures and government policies, differ between countries (Jamieson 1991; Kraan *et al.* 1991) but the general pattern is broadly similar across Western societies (as well as many other parts of the world). In some senses, this process can be seen to have dispersed elements of the institution into the community, as the older person (now defined as a service user or consumer) becomes identified with the receipt of care, and the space and time of home become controlled by others.

At the same time, home and community care differs from residential care in that it is not *total* institution. While some aspects of the person's life may become subject to control by others, and social horizons become more limited by mobility, it may be argued that other aspects of life break through the boundaries of home and merge with the care régime. Family, friends and neighbours may form part of the care process, but they may also visit or be visited. Other social activities may also continue, even though 'care' may include arrangements for transport to enable this to happen, holidays may be taken and so on. The degree of personalisation and individuality in daily life is thus greater than either residential care or the congregate life of retirement communities. Usage of home and community care services in later life represents a rite of passage that can be integrated with a sense of 'normality'. In contrast, the move into residential care, or even into an exclusive retirement community, constitutes a sharp break that emphasises the transition to a new social status which is, to some extent, socially devalued. Yet, for the older person who stays in their own home, the maintenance of 'normality' is not inevitable and will only be achieved within a discourse that is oppositional to the ageism endemic in current ideas and practices. So, to conclude, I will examine the possibility of constructing old age in a way which differs from the ageism that has been described above.

Body/space/image in oppositional discourse

This chapter has focused on three forms of living arrangements in later life that relate to aspects of 'care' and hence dependency. It has been argued that in different ways, each of these forms represents a construction of old age through the linking of body, space and image. The old body, defined in terms of decline and decay, is managed in certain types of spaces, congregated and segregated, thus building and maintaining an image of old age as 'other to' and with implications of 'less than' the mainstream of adult life. The parallels of the distinction young/old and male/female (Harper 1997) have also been observed.

To what extent should later life simply be understood in these terms? Is it feasible or desirable to oppose the inherent stigmatisation of 'old age' inherent in such an understanding, with its consequences of marginalisation and denigration? Fennell *et al.* (1988), Biggs (1993) and Bytheway (1995) have each argued that to disentangle ageing and later life from oppression and discrimination is not an easy task: ageism is endemic to contemporary Western culture. Blakemore and Boneham (1994) and Harper (1997) have shown that ageism is interwoven with other social divisions (such as race, ethnicity and gender) (cf. Bytheway 1995: 117–118). Indeed, it may be the case that people of the same generation *choose* to share time and activity with each other in particular spaces; the common life experiences of shared biography can provide a basis for companionship. Despite this, it must be emphasised that, in an ageist society, such choice is often imposed by younger people.

An oppositional discourse would be one that enabled the construction of identity around all the facets of a person, in context, and did not reify particular features, of which age is a key instance. As Bytheway notes (1995: 128), this is not the same as 'thinking positively' about age: that can be equally patronising. Rather, it requires a recognition of the complexities and contradictions of reality and, most important, a shift from 'us/them' discourse to an inclusive view that connects people at different stages of the life-course as part of the same society. The capacity to shape as well as be shaped by the social terrain in which we live is an important ingredient in this alternative construction of ageing (Tout 1995).

Yet, in contemporary circumstances, the imposition of negative stereotypes by more powerful 'others' serves to create discriminatory rites of passage in ageing. The denial of older people's capacities, in popular discourse as well as in that of the professions, policy makers or academics, means that to become 'old' is to acquire a spoiled identity (Goffman 1971), even where a physical 'passage' is not clearly defined, as in home and community care contexts. Chronological age in itself carries with it implications for social identity that are embodied through the popular association of later life with dependence and the loss of faculties.

The alternative to this, which might be the mark of an oppositional discourse, is not an idealised romantic 'reconstruction' of the valued role of 'elder' or the (patronising) celebration of the 'disco-granny' or the 'parachuting pensioner', but the incorporation into the mainstream social world of *all* people, irrespective of age, including those (of *us*) who require assistance in daily living and so choose to use home and community care services or who, for whatever reason, wish to live in congregate housing. It would also provide important opportunities for inter-generational relationships that were based on diverse social roles that include older people as providers as well as recipients of care, and as contributing to social and cultural life (Brabazon and Disch 1997).

As I write this chapter I am not yet 'old' but I am 'ageing'. Each of us may say with Paul Simon (1970) 'I am older than I once was, but I'm younger than I'll be, and that's not unusual – no it isn't strange'. This process involves many rites of passage, yet none should define our entire identities: each may contribute to who we are and who we are becoming, but this is a dynamic process. In contrast to J.B. Priestley's critical examination of his sense of the ageing self, it is feasible, even desirable, that we should seek to create a realistic view of later life that is neither patronisingly idealistic or oppressively negative. Perceptions of incongruence between how older people feel and their images of old age appear to be common (cf. Blythe 1979; Featherstone and Hepworth 1989; Hazan 1994). So the issue would seem to lie more in the cultural and social differentiation of youth and age, and the rites of passage between them, than in any 'essence' of later life as such. A discourse oppositional to ageism, therefore, must seek to 'normalise' old age (Bytheway 1995: 128) in such a way that the diversity of experience is regarded both realistically and positively as an *ordinary* part of mainstream adult life.

Notes

1 There is an abiding myth in Western society that 'families no longer care for their elders'. Research evidence from Australia (McDonald 1997), the UK (Finch 1989) and the USA (Newman 1988) points to the extensive family networks in which older people are located. The relatively higher proportions of institutional care in Western countries compared to other parts of the world do not arise because 'the family' has

'failed' older people. This phenomenon stems from shifts in demographic profiles of these countries, which means that a much larger number of people are surviving longer into that age range where care and support is more likely to be needed (Peace *et al.* 1997). In these societies, it was never culturally normative for older people to share a dwelling with younger relatives (Jefferys and Thane 1989: 15). So when family care is provided, it is more usually in the older person's own home. Residential care therefore is not a replacement for a tradition of older people moving to live with younger relatives.

2 Some critical gerontologists have argued that the focus on need and negativity should give way to a more rounded view of all aspects of 'normal' old age (Fennell *et al.* 1988). To a degree I concur with this view, although it is also plausible to conclude that, unless one is focusing on the differences between later life and earlier stages of the life-course, it may be impertinent to talk of old age at all. Yet it is the common experience of the construction of later life and old age as different, which appears to some extent in all cultures (Minois 1989; Keith 1992), that commands our analytical attention. For these reasons, despite the clear issues raised in relation to imaging and identity, discussions of 'normal' old age invariably take in questions of need and care in social context as well as 'difference' in social statuses and roles (Hugman 1994: 19–20).

References

Allen, I., Hogg, D. and Peace, S. (1992) *Elderly People: Choice, Participation and Satisfaction*, London: Policy Studies Institute.

Anderson, R. (1992) 'Health and community care', in L. Davies (ed.) *The Coming of Age in Europe*, London: Age Concern England, 63–84.

Bartlett, H. (1993) *Nursing Homes for Elderly People: Questions of Quality and Policy*, Chur, Switzerland: Harwood Academic Publishers.

Biggs, S. (1993) *Understanding Ageing*, Buckingham: Open University Press.

Blakemore, K. and Boneham, M. (1994) *Age, Race and Ethnicity*, Buckingham: Open University Press.

Blythe, R. (1979) *The View in Winter*, London: Allen Lane.

Bond, J., Briggs, R. and Coleman, P. (1993) 'The study of ageing', in J. Bond, P. Coleman and S. Peace (eds) *Ageing in Society*, London: Sage Publications, 19–52.

Booth, T. (1985) *Home Truths: Old People's Homes and the Outcome of Care*, Aldershot: Gower.

Brabazon, K. and Disch, R. (eds) (1997) *Intergenerational Approaches in Aging: Implications for Education, Policy and Practice*, New York: The Haworth Press.

Briggs, P. (1993) 'Biological ageing', in J. Bond, P. Coleman and S. Peace (eds) *Ageing in Society*, London: Sage Publications, 53–67.

Bytheway, B. (1995) *Ageism*, Buckingham: Open University Press.

California Association of Homes and Services for the Aging (1997) *Senior Sites*, http://www.seniorsites.com (downloaded 25 February).

Clough, R. (1981) *Old Age Homes*, London: George Allen and Unwin.

Esping-Andersen, G. (1990) *The Three Worlds of Welfare Capitalism*, Cambridge: Polity Press.

Featherstone, M. and Hepworth, M. (1989) 'Ageing and old age: reflections on the post-modern life course', in B. Bytheway, T. Keil, P. Allatt and A. Bryman (eds) *Becoming and Being Old*, London: Sage Publications, 143–157.

Fennell, G., Phillipson, C. and Evers, H. (1988) *The Sociology of Old Age*, Milton Keynes: Open University Press.

Finch, J. (1989) *Family Obligations and Social Change*, Cambridge: Polity Press.

Foucault, M. (1977) *Discipline and Punish*, Harmondsworth: Peregrine.

Goffman, E. (1968) *Asylums*, Harmondsworth: Penguin.

—— (1971) *The Presentation of the Self in Everyday Life*, Harmondsworth: Penguin.

Goodman, R.J. and Smith, D.G. (1992) *Retirement Facilities: Planning, Design and Marketing*, New York: Whitney Library of Design.

Gordon, R.A. (1993) *Developing Retirement Communities*, Vol. 1, New York: John Wiley and Sons.

Green, T. (1993) 'Institutional care, community services and the family', in P.G. Koopman-Boyden (ed.) *New Zealand's Ageing Society: the Implications*, Wellington: Daphne Brasell Associates Press, 149–186.

Greer, G. (1991) *The Change: Women, Ageing and the Menopause*, London: Hamish Hamilton.

Grundy, E. and Harrop, A. (1992) 'Demographic aspects of ageing in Europe', in L. Davies (ed.) *The Coming of Age in Europe*, London: Age Concern England, 14–37.

Harper, S. (1997) 'Constructing later life/constructing the body: some thoughts from feminist theory', in A. Jamieson, S. Harper and C. Victor (eds) *Critical Approaches to Ageing and Later Life*, Buckingham: Open University Press, 160–172.

Hazan, H. (1994) *Old Age: Constructions and Deconstructions*, Melbourne: Cambridge University Press.

Henrard, J.-C., Ankri, J. and Isnard, M.-C. (1991) 'Home-care services in France', in A. Jamieson (ed.) *Home Care Services for Older People in Europe: A Comparison of Policies and Practices*, Oxford: Oxford University Press, 99–117.

Hugman, R. (1991) *Power in Caring Professions*, Basingstoke: Macmillan.

—— (1994) *Ageing and the Care of Older People in Europe*, Basingstoke: Macmillan.

—— (1996) 'Health and welfare policy and older people in Europe', *Health Care in Later Life*, 1, 4: 211–222.

Jamieson, A. (1991) 'Home care in Europe: background and aims', in A. Jamieson (ed.) *Home Care for Older People in Europe: A Comparison of Policies and Practices*, Oxford: Oxford University Press, 3–12.

Jefferys, M. and Thane, P. (1989) 'An ageing society and ageing people', in M. Jefferys (ed.) *Growing Old in the Twentieth Century*, London: Routledge, 1–18.

Keith, J. (1992) 'Care taking in cultural context: anthropological queries', in H. Kendig, A. Hashimoto and L.C. Coppard (eds) *Family Support for the Elderly: The International Experience*, Oxford: Oxford University Press/World Health Organization, 15–30.

Kraan, R.J., Baldock, J., Davies, B., Evers, A., Johansson, L., Knapen, M., Thorslund, M. and Tunissen, C. (1991) *Care for the Elderly: Significant Innovations in Three European Countries*, Frankfurt am Main/Boulder, Colorado: Campous Verlag/Westview Press.

Laslett, P. (1989) *A Fresh Map of Life: The Emergence of the Third Age*, London: Weidenfeld and Nicholson.

Latimer, J. (1997) 'Figuring identities: images of care relationships with older people', in A. Jamieson, S. Harper and C. Victor (eds) *Critical Approaches to Ageing and Later Life*, Buckingham: Open University Press, 143–159.

Laws, G. (1995) 'Embodiment and emplacement: identities, representation and landscape in Sun City retirement communities', *International Journal of Aging and Human Development* 40, 4: 253–280.

—— (1997) 'Spatiality and age relations', in A. Jamieson, S. Harper and C. Victor (eds) *Critical Approaches to Ageing and Later Life*, Buckingham: Open University Press, 90–101.

McCallum, J. (1997) 'Health and ageing', in A. Borowski, S. Encel and E. Ozanne (eds) *Ageing and Social Policy in Australia*, Melbourne: Cambridge University Press, 54–73.

McCallum, J. and Geiselhart, K. (1996) *Australia's New Aged: Issues for Young and Old*, Sydney: Allen and Unwin.

McDonald, P. (1997) 'Older people and their families: issues for policy', in A. Borowski, S. Encel and E. Ozanne (eds) *Ageing and Social Policy in Australia*, Melbourne: Cambridge University Press, 194–210.

Minois, G. (1989) *History of Old Age*, Cambridge: Polity Press.

Newman, S. (1988) *Worlds Apart? Long-term Care in Australia and the United States*, New York: Haworth Press.

Office of Population Censuses and Surveys (1988) *Disability Survey*, Vols I and II, London: HMSO.

Ozanne, E. (1997) 'Ageing citizens, the state and social policy', in A. Borowski, S. Encel and E. Ozanne (eds) *Ageing and Social Policy in Australia*, Melbourne: Cambridge University Press, 233–248.

Peace, S., Kellaher, L. and Willcocks, D. (1997) *Re-evaluating Residential Care*, Buckingham: Open University Press.

Phillipson, C. (1982) *Capitalism and the Construction of Old Age*, London: Macmillan.

Sax, S. (1993) *Ageing and Public Policy in Australia*, Sydney: Allen and Unwin.

Scull, A. (1979) *Museums of Madness*, London: Allen Lane.

—— (1984) *Decarceration*, 2nd edn, Cambridge: Polity Press.

Simon, P. (1970) 'The boxer', from *Simon and Garfunkel – The Concert in Central Park*, New York: CBS Records.

Sinclair, I., Parker, R., Leat, D. and Williams, J. (1990) *The Kaleidoscope of Care: A Review of Research on Welfare Provision for Elderly People*, London: HMSO.

Sontag, S. (1978) 'The double standard of ageing', in V. Carver and P. Liddiard (eds) *An Ageing Population: Raeder and Sourcebook*, Sevenoaks: Hodder and Stoughton, 72–80.

Thurow, L. (1996) *The Future of Capitalism*, Sydney: Allen and Unwin.

Toner, J.A. and Kutscher, A.H. (1993) 'The evolving continuum of long-term care', in J.A. Toner, L.M. Tepper and B. Greenfield (eds) *Long-term Care: Management, Scope and Practical Issues*, Philadelphia: The Charles Press Publishers, 1–24.

Torres-Gil, F.M. (1992) *The New Aging: Politics and Change in America*, Westport, Connecticut: Auburn House.

Tout, K. (1995) 'An ageing perspective on empowerment', in D. Thursz, C. Nusberg and J. Parther (eds) *Empowering Older People: An International Approach*, Westport, Connecticut: Auburn House, 3–35.

Ungerson, C. (1983) 'Why do women care?', in J. Finch and D. Groves (eds) *A Labour of Love: Women, Work and Caring*, London: Routledge and Kegan Paul, 31–49.

Van Gennep, A. (1960) *The Rites of Passage*, trans. M.B. Vizedom and G.L. Caffee, London: Routledge and Kegan Paul. First published in 1909, *Les Rites de Passage*, Paris: Noury.

Willcocks, D., Peace, S. and Kellaher, L. (1987) *Private Lives in Public Places*, London: Tavistock.

Williams, S. (1986) 'Long-term care alternatives: continuing care retirement communities', in I.A. Morrison, R. Bennett, S. Frisch and B.J. Gurland (eds) *Continuing Care Retirement Communities: Political, Social and Financial Issues*, New York: Haworth Press, 15–34.

Wolfensberger, W. (1972) *Normalization*, Toronto: National Institute of Mental Retardation.

Woodward, K. (1991) *Aging and its Discontents*, Bloomington: Indiana University Press.

13

THE TRANSITION INTO ELDERCARE

An uncelebrated passage

Bonnie C. Hallman (Canada)

Family life in patriarchal Western cultures is very different for men and for women. This reflects the way that family members fulfil different roles. Furthermore, women and men experience the evolution of family life over time often in quite different ways. In any culture, the sorts of social interactions in which men and women engage with their families, the time invested in those activities and their time and space 'context' (where the activities occur and over what distance) all reflect the gendered values and norms for family life within that culture (Massey 1994).

This chapter advances two ideas. The first is that the provision of assistance to elderly relatives ('eldercare'; assistance given to relatives over the age of 65) occurs in a gendered time–space context. The second is that many people, particularly women, find that their new role of eldercare provider is a troublesome and largely uncelebrated passage into a new life phase. In the process of this transition, changes may well occur in the temporal and spatial context – or 'personal geographies' – of both the family care-giver and the elderly relative(s) receiving assistance. This reflects the fact that family-based assistance for elderly relatives occurs within a highly gendered social context. Empirical studies consistently identify two-thirds to three-quarters of primary care-givers as female family members, predominantly wives, daughters and daughters-in-law (Abel 1989; Brody 1990; Chappell 1991). Furthermore, the performance of care-giving tasks tends to follow traditional gender divisions of household labour. Women predominate in personal care, such as feeding and bathing, and in help with tasks such as meal preparation and housekeeping, assistance that tends to be more routine and intimate in nature than the traditionally male activities such as home maintenance and financial assistance (Finley 1989; Kaye and Applegate 1990).

This need for eldercare is increasing. In all of the developed world, populations are ageing. In the USA alone, 1994 census data put the over-65 population at 33 million, and this number will rise to nearly 70 million in less than thirty years (Berman 1996). Many of these elderly persons will live to advanced ages, but they are more likely to live with illness and disability. At the same time, more of them will live at a distance from their children, some in different cities, but some in different regions entirely (Googins 1991; Berman 1996; Foot and Stoffman 1996).

This 'geography of eldercare' is, however, not clearly understood. Research in human geography has demonstrated the tenacity of gender roles and their profound effect on the spatial behaviour of women (Duffy *et al.* 1989; Katz and Monk 1993; Hanson and Pratt 1995), with emphasis on effects on women's labour force participation. When attention has been paid to family and household responsibilities the emphasis has been on child care (Dyck 1990; Gregson and Lowe 1995) but if, in addition, we turn our attention to care-giving relationships in older families, we further our knowledge of the effects of time–space relationships on social behaviour as well as our understanding of the transitions associated with ageing families.

Transitions into gendered geographies of eldercare

In this chapter, I aim to accomplish two main tasks. First, I show the influence of gendered time–space relationships on the extent and frequency of family care-giving to elderly relatives in need of some assistance. I use data from a case study of Canadian employed family care-givers to elderly relatives. The hypothesis behind this analysis is twofold: first, that the eldercare behaviour of men and women will reflect different relationships to time and space (Rose 1993; Davies 1994), and second that these gender differences will translate into demonstrably different patterns of eldercare provision and negotiation: *gendered geographies of eldercare* (Hallman 1997; Joseph and Hallman 1998). These consist of both the time-space paths created through the regular movement of care-givers, in time and over space, in their travels to their elderly relatives' homes in order to provide assistance *and* any adjustments made in their locations. Therefore, particular attention is placed on *residential relocation* as a uniquely geographical means of modifying the context of family-based care-giving to elderly relatives (Joseph and Hallman, 1996; Hallman 1997).

Second, I will introduce the concept that residential relocation can be understood as a watershed moment, perhaps even a 'rite of passage', in the transition into the new role of eldercare provider. It can be argued that the adoption of this role marks the beginning of a new life stage. Other major life-stage transitions, such as adolescence, marriage and the birth of a child, are marked, even in modern societies noted for their lack of ritual, by some form of ceremony or celebration in which many members of a group share in recognising the individual's status change. These ceremonies or 'rites of passage' (e.g. bar mitzvah, wedding ceremony, baby shower, baptism, naming ceremony) often act as cushioning through the transition experience, and can assist in the establishment of some sort of new equilibrium based on the new social role(s) into which an individual has moved (Chapple and Coon 1942; van Gennep 1960). This discussion will focus first on changes in time–space relationships, especially the rate of residential relocation, as 'markers' for the new social role of 'eldercare provider'. This will be followed by a discussion of these adjustments to personal eldercare geographies as points of transition, where the responsibilities and commitments associated with care-giving can at least be prepared for, if not celebrated. These changes are in lieu of merely 'making do' and perhaps being overwhelmed by the pressures and stresses of trying to balance employment, family and eldercare (Abel and Nelson 1990).

This chapter has three sections. The first briefly introduces some of the major themes in the social science literature on the space and time context of family care-giving. The focus here is twofold: the gendered experience of time and space as it shapes the social and spatial interactions of family care-giving to elderly relatives; and the utility of thinking of this experience as a geographical hallmark of the, largely unheralded, transition into a new social role or life stage.

The second section presents key findings from a case study of employed family care-givers to their elderly relatives who are still living independently in their communities. These are not older people living in institutions such as nursing homes or homes for the aged. Time-distance in minutes is the unit of measure used for the distance between care-givers and their older relatives (time–space context). This measure recognises that in our hurried culture time-management and time-budgeting are central to families striving to balance family needs with paid work and other community responsibilities (Googins 1991; Daly 1996; Joseph and Hallman 1996). I will focus on three areas: average hours of eldercare provision; the frequency of assistance for specific care tasks; and how often family eldercare providers consider, and implement, residential relocation as a coping strategy.

The case study analysis is augmented by care-givers' comments on the effects of time and distance on their interactions with their families, and on their provision of care and assistance to their older relatives. Other care-giver comments address residential relocation and its place in the transition into eldercare. The use of care-givers' own words allows us to 'hear' the challenges and trials of this adjustment to a new status and set of responsibilities from people in the midst of the transition into the role of family eldercare provider.

The third and last section offers a discussion of the potential insight gained into the transition into eldercare from a time–space perspective, and from viewing the adjustments of personal eldercare geographies as hallmarks of a life-course transition.

Space, gender and time: family eldercare in context

Living nearby is widely recognised as an integral factor in determining the extent and nature of participation in the family-work of eldercare, particularly parent-care (Stueve and O'Donnell 1989; Chappell 1991; Stoller *et al.* 1992). Indeed, residential proximity is the first and strongest criterion of availability and interaction between family members (Litwak and Kulis 1987). Nearby children of elderly parents needing some help are more likely to be called upon to assist than are those living further away. This, of necessity, involves a more extensive time and energy commitment than more distant siblings can manage. These more distant family members may help financially, assist with short-term crises or offer emotional support to the elderly relative and/or a more proximate family care-giver, but sheer distance may prevent a more equitable sharing of care-giving responsibilities (Schoonover *et al.* 1988). A distance barrier, therefore, must be overcome in order to provide eldercare (Litwak and Kulis 1987).

Some types of care-giving assistance, especially Activities of Daily Living (ADLs, *e.g.* feeding, bathing and toileting) which demand frequent and face-to-face contact, will be more keenly hindered by this barrier. Other forms of assistance, such as transportation and home maintenance, are done on a much less frequent basis and thus the limiting effects of distance may be minimal by comparison (Silverstein and Litwak 1993).

Care-giver gender also plays a role in adult children's response to elderly parents in need, especially when we investigate women's care-giving responses over distance. Patriarchal societal expectations, stemming from the traditional female 'nurturer' role, may have an influence on behaviour that conflicts with highly gendered limitations on spatial behaviour faced by many women, such as restricted access to personal transportation (Katz and Monk 1993; Hanson and Pratt 1995), and prioritisation of family concerns over employment outside of the home or personal time (Duffy *et al.* 1989; Kobayashi *et al.* 1994).

Relationships that involve care-giving bring into play a set of competing time dynamics. Most institutional forms of care-giving are shaped by the same linear time form that governs all productive (traditionally male) activity. Family-based care operates under a different conceptualisation of time. Care-giving inherently requires process time; needs are often not predictable and a plurality of caring activities occur within any given situation or experience (Davies 1994). Thus, it can be argued that, by its very nature, the traditionally female labour of family care has a different relationship to time – that it is grounded differentially in a gendered time–space of eldercare. That is, gender relationships and roles, especially those aspects associated with the care of family members, structure men's and women's relationships to time and to space, and thus their paths through time and space quite differently (Rose 1993).

This gendered relationship to time is made tangible in relationships to space as time-distance, which can be measured in minutes of travel time. Gendered time–space relationships are also reflected in eldercare activities, such as the average time committed to eldercare assistance, including the time spent travelling in order to provide eldercare tasks, and the use of residential relocation. These three areas are explored in the following presentation of key results from a case study of employed Canadian men and women assisting at least one elderly family member.

The Work & Family Eldercare case study

Data for this case study of employed men and women eldercare providers are drawn from surveys conducted in 1991 and 1995 by the Work and Eldercare Research Group of the Canadian Aging Research Network (CARNET).[1] The 1991 Work & Family Survey involved the participation of employees from eight Canadian organisations within the industrial, services and public sectors. Survey questionnaires were distributed in the work place and returned by mail, for a response rate of 53 per cent (5,121 of the 9,693 distributed). Every Canadian province is represented with the exception of Prince Edward Island; Ontario is over-represented at over 60 per cent of the respondents. The 1995 Work & Eldercare Survey was a follow-up involving a sub-set of 250 employed eldercare providers. This survey provides a rich source of opinion and comment on the experience, positive and negative, of providing assistance to elderly relatives; the personal dimension of eldercare.

Of the men and women who participated in the 1991 Work & Family Survey, 1,149 reported providing assistance to an elderly relative (over the age of 65) living independently in their communities (n = 836: 72.3 per cent) or helping two elderly relatives living at the same location (n = 313: 27.2 per cent) at least once during the six months prior to participation in the Work & Family Survey. Most were between the ages of 36 and 55 (85 per cent) and married or in common-law relationships (84 per cent). More women than men reported part-time (>35 hours per week) employment (19 per cent compared with 1.6 per cent).

As stated earlier, travel time is the unit of measurement for the distance between care-givers and their elderly relatives. All care-givers were put into one of three time-distance categories: 1–30 minutes (n = 703), 31–120 minutes (n = 297) and more than 120 minutes (n = 149).[2] These categories correspond roughly to three types of travel: *short distance* trips which could reasonably be done on a daily or near-daily basis; longer but still *moderate distance* trips which could be made in a single day or over a weekend; and *long distance* travel which would necessitate extended time away from paid work and other family commitments.

Time devoted to eldercare

The data in Table 13.1 shows that time-distance significantly affects the average hours devoted per week to eldercare by the people in the case study. Hours of care drop from a high of 4.27 hours on average among family care-givers living within 30 minutes of their elderly relative, to a low of 3.06 hours per week among long-distance (>120 minutes of travel) eldercare providers. This follows the notion of *distance-decay*, which states that the interaction between two objects or places decreases as the distance (however measured) between them increases (Knox and Marston 1998). However, the effect of time-distance on hours spent providing eldercare is not the same for men and women. In all three time-distance categories, women provide significantly more hours of care than do their male counterparts; on average more than an hour of eldercare assistance per week. This is particularly evident among female care-givers 31-120 minutes from their elderly relative. They report providing on average 4.63 hours of eldercare per week, almost identical to the 4.66 hours reported by women care-givers who live within 30 minutes of their relative. In contrast, male care-givers seem to sharply reduce their time spent in eldercare once the time-distance exceeds 30 minutes (3.55 to 2.75, see Table 13.1).

The data for all care-givers, therefore, support the general notion of distance-decay. Like other forms of interaction, time spent helping elderly relatives decreases as the distance increases between family care-giver and older relative. However, care-giver gender significantly differentiates this pattern. Women care-givers are less likely to limit the time they spend on eldercare than their male counterparts. Time-distance does not significantly reduce women's eldercare time-commitment until the time-distance exceeds two hours. Rather, gender norms regarding family care for older family members appear to underwrite a *distance-defying* commitment to eldercare among the women in this case study (see also Hallman and Joseph, forthcoming).

This traditional role effect was investigated a little further through disaggregation according to the presence of minor children (under 18 years old) in the home. The data show that the significant reduction in time spent in eldercare among all care-givers is largely driven by the behaviour of care-givers with minors (Table 13.1). Once time-distance exceeds half an hour, the average time spent on care-giving drops off quickly among those with young children (Table 13.1) and the data show that this is driven primarily by male behaviour, in that long-distance male care-givers with young children at home provide significantly

Table 13.1 Average hours per week of eldercare, by time-distance, care-giver gender and responsibility for children

Care-giver groups	*Time-distances (in minutes)*						
	1–30		*31–120*		*>120*		*Analysis of variance*
	Mean	*(n)*	*Mean*	*(n)*	*Mean*	*(n)*	*(p)*
All care-givers	4.27	(703)	3.84	(297)	3.06	(149)	0.018
Male care-givers	3.55	(253)	2.75	(121)	2.31	(70)	0.029
Female care-givers	4.66	(446)	4.63	(176)	3.74	(78)	0.353
All care-givers: children <18	4.49	(394)	3.60	(177)	3.13	(77)	0.021
Male care-givers: children <18	3.66	(147)	2.70	(75)	2.07	(35)	0.066
Female care-givers: children <18	4.92	(244)	4.27	(102)	4.27	(42)	0.468
All care-givers: no children <18	4.02	(309)	4.20	(120)	2.96	(72)	0.230
Male care-givers: no children <18	3.43	(102)	2.83	(46)	2.70	(35)	0.447
Female care-givers: no children <18	4.37	(202)	5.16	(74)	3.15	(36)	0.211

Source: Hallman and Joseph, forthcoming

Note:
A significance level of 0.05 is used as the benchmark for statistical significance, with the exception of the Analysis of Variance result for male care-givers with children under 18. Here, a significance level of .10 is used

fewer hours of eldercare assistance (less than half the number of hours) than do their similarly situated female counterparts (Table 13.1).

These results corroborate evidence in academic and popular literature of working women's tendency to add additional family responsibilities on to existing work and family commitments rather than delegating such responsibilities to other family members or to formal services (Abel and Nelson 1990; Friedan 1991; Glendon 1997). In summary, the data illustrate the influence of gender norms in the responsibility for care-giving, both for the elderly and for minor children. Women in this sample did not significantly reduce their time-commitment to eldercare when they lived at a distance from their older relatives. Responsibility for child care also does not appear to cause a significant reduction in the time spent on care to elderly relatives. Only male care-givers significantly reduced the hours spent on eldercare as the distance between themselves and their relatives increased.

Adding new facets to the traditional female role of care-giver can be understood as a common part of the transition into eldercare. While this can be seen as the continuation of a role an individual woman may have played since the birth of a first child, the switch in focus to the older generation marks eldercare as quite different, as does the nature of the development of the care-giving activity. Caring for children normatively involved increasing independence and the development at some point of a relationship of equals, while eldercare is associated with increasing levels of need and dependence, ending only with death (Abel and Nelson 1990; Googins 1991).

Time-distances and assistance frequency

The average time-distance at which a particular type of assistance task is provided appears to be related predictably to the frequency at which it is provided (Table 13.2). As might be expected, high-frequency assistance (daily or several times a week) is associated with travelling lower average distances. This trend is statistically significant for the three instrumental activities (IADL: meal preparation, shopping and home maintenance), but not for the more intensive and personal activity of bathing (an ADL activity: Table 13.2) (see also Joseph and Hallman 1998). However, considering the frailty of health associated with the need for bathing assistance, logic suggests that care-givers prepared to provide this level of assistance will travel further to ensure that such personal needs are met by family.

Shopping appears to be distinctive from the other forms of eldercare assistance considered here. For both male and female respondents, there is a clear, and very similar, relationship between the frequency at which shopping help is given and the average distances men and women care-givers will travel in order to provide that assistance. In addition, male care-givers in this case study seem less willing than females to overcome distance barriers to provide frequent home maintenance or meal preparation. In addition, women care-givers are almost always more willing to travel further to provide assistance at any given frequency than are men. This appears to be especially so for high-frequency care, although interpretation must be guarded due to the small number of male care-givers involved. Nevertheless, for all tasks but shopping, women care-givers report on average travelling longer to provide assistance (Table 13.2).

Do changes in the personal geographies of care-givers to elderly relatives translate into adjustments, specifically residential relocation? And can moving be understood as another step in the passage into the role of family eldercare provider? These questions are addressed in the next section.

Table 13.2 Average time-distances travelled to provide care in four tasks, by assistance frequency

Frequency of assistance[a]	*All care-givers*		*Male care-givers*		*Female care-givers*	
	Mean	*(n)*	*Mean*	*(n)*	*Mean*	*(n)*
ADL[b]: bathing						
High	35.1	(15)	23.8	(4)	39.3	(121)
Medium	53.8	(35)	127.8	(4)	46.6	(29)
Low	66.9	(56)	52.0	(5)	68.4	(51)
P	0.375		0.305		0.357	
IADL[c]: meal preparation						
High	24.4	(23)	19.0	(5)	25.8	(18)
Medium	46.7	(95)	45.8	(20)	47.9	(73)
Low	119.5	(208)	127.7	(58)	101.5	(149)
P	0.013		0.045		0.122	
IADL: shopping						
High	16.6	(41)	17.3	(9)	16.4	(32)
Medium	26.8	(253)	27.9	(81)	26.5	(169)
Low	86.6	(352)	85.4	(119)	86.6	(232)
P	0.000		0.003		0.001	
IADL: home maintenance						
High	20.4	(21)	16.2	(12)	26.1	(9)
Medium	36.7	(171)	32.2	(68)	40.1	(101)
Low	85.4	(433)	79.9	(202)	90.5	(229)
P	0.010		0.006		0.180	

Source: Hallman 1997

Notes:

a High frequency equals daily or several times a week; medium frequency equals weekly or several times a month; low frequency equals once a month or less

b ADL: activity of daily living, such as feeding and toileting

c IADL: instrumental activities of daily living, such as preparing meals, housework and so on

Residential relocation: (re)negotiating eldercare geographies

The distance between care-giver and elderly relative – the 'time–space path' of eldercare – is the site of individual and very geographical adjustments in the negotiation of eldercare responsibilities. Three types of relocation are considered: (1) urging or arranging for the elderly relative to move closer to the care-giver; (2) urging or arranging for the elderly relative to move into the care-giver's home; and (3) the consideration of institutionalisation. All data reported here are for moves considered or arranged within a six-month period prior to care-givers' participation in the CARNET Work & Family Survey.

Table 13.3 Residential relocation of the elderly relative, by care-giver gender

Residential relocation care-givers options	*All care-givers*		*Male care-givers*		*Female care-givers*	
	%	(*n*)	%	(*n*)	%	(*n*)
Urged/arranged for relative to move closer	12.4	(140)	9.7	(43)	14.1	(97)
				$\chi^2 = 4.77$ $p = 0.029$		
Urged/arranged for relative to move into care-giver's home	6.4	(72)	4.5	(20)	7.6	(52)
				$\chi^2 = 4.15$ $p = 0.042$		
Looked into institutional care for the relative	13.5	(152)	13.8	(61)	13.2	(91)
				$\chi^2 = 0.079$ $p = 0.778$		

Table 13.3 shows us the effect of care-giver gender, separate from the effects of care-giver location (time-distance from their elderly relative) on the consideration of the three residential relocation options. For the first two options (moving an elderly relative closer, or into the family care-giver's home) female care-givers are significantly more likely to report these options than their male counterparts.

When we focus on the distance between care-giver and elder (Table 13.4), the incidence of consideration or action generally increases for all residential relocation options as the time-distance (minutes of travel time) between care-giver and elder increases, and significantly so for the *urged/arranged to move elder closer* option.[3]

Adding the effects of care-giver gender (Table 13.4), the significant relationship between the *urging or arranging for an elderly relative to move closer* option, and time-distance, is revealed to be largely the result of the decisions of female care-givers. Indeed, more than 20 per cent of female, long-distance eldercare providers (more than 120 minutes travel time, one way, to the relative they help) report urging, or arranging for, this residential relocation option within the six-month pre-survey period. Among the men in the case study, none of the residential relocation options are very popular, and consideration or use of residential relocation does not increase when they live further away from their elderly relatives (Table 13.4).

Making time, and space, for eldercare

Comments made by the eldercare providers who responded to the Work & Eldercare survey address, some quite eloquently, the issues of *making time for eldercare*, and of making changes in their lives as they proceed through the

Table 13.4 Residential relocation of the elderly relative, by time-distance and care-giver gender

Residential relocation options	*Time distances (in minutes)*	*All care–givers*		*Male care–givers*		*Female care–givers*	
		%	*(n)*	*%*	*(n)*	*%*	*(n)*
Urged/arranged for relative to move closer	1–30	9.8	(69)	9.1	(23)	10.4	(46)
	31–120	15.8	(47)	9.9	(12)	19.9	(35)
	>120	16.1	(24)	11.4	(8)	20.5	(16)
		χ^2=10.0 *p*=0.007		χ^2=0.344 *p*=0.842		χ^2=14.1 *p*=0.001	
Urged/arranged for relative to move into care-giver's home	1–30	5.5	(39)	4.4	(11)	6.3	(28)
	31-120	7.4	(22)	2.5	(3)	10.8	(19)
	>120	7.4	(11)	8.6	(6)	6.4	(5)
		χ^2=1.7 *p*=0.426		invalid		χ^2=4.36 *p*=0.113	
Looked into institutional care for the relative	1–30	13.1	(92)	14.2	(36)	12.6	(56)
	31-120	14.5	(43)	14.0	(26)	14.8	(26)
	>120	11.4	(17)	11.4	(8)	11.5	(9)
		χ^2 0.968 p=0.616		χ^2=0.405 p=0.817		χ^2=0.950 p=0.622	

Notes: Adapted from Hallman 1997

transition to eldercare. These 'voices' help in the interpretation of the data presented above, by illustrating the ways that time and space structure the eldercare transition experiences of individual family care-givers.

The following two responses, for example, illustrate an awareness among some care-givers of the strain of fitting it all in as they add care for older relatives to their other work and family responsibilities. These care-givers particularly emphasise impacts on their marital relationships. A 42-year-old woman who provides an average of eight hours of eldercare per week to her elderly mother said: 'My husband and I don't spend as much weekend time together as we'd like, and seldom go away for a weekend. My mother isn't too happy if we go away for a weekend either'.

Another female care-giver, 38 years old and providing an average of five hours of assistance per week to her elderly parents, and who has two minor children at home, speaks of similar feelings: 'The burden of caring for parents and children leaves one (or a couple) with no time for themselves. This is a big problem and you begin to resent the constant demands of relatives'.

These sorts of comments show that while women eldercare providers may be spending significantly more time on eldercare provision (Table 13.1), and providing more frequent assistance for certain tasks (Table 13.2), than their male counterparts, this is not considered unproblematic. As these two women indicate, there are costs involved to their personal relationships, particularly trade-offs between their marital relationships and their relationships with their parents. This problem of competing family demands on the care-giver's personal time is approached in a slightly different way by another care-giver. This 48-year-old woman, who provides an average of thirteen hours of assistance per week to her elderly parents, was asked about potential eldercare community service improvements. She stated the need for: 'Some assistance for weekends, evenings – a period of time of continuous care so my husband and I could get away for a couple of days'. This woman describes a need for community service support for herself in order to find a better balance between the needs of her elderly parents and the need for time to spend on maintaining her marital relationship. As other authors have shown (Abel and Nelson 1990; Chappell 1991) this conflict between relationships, as women in particular attempt to meet their commitments to several family relationships, is an all too common feature of the transition into eldercare.

The comments made by these same men and women about the impacts of residential relocation on their experience of providing eldercare further demonstrate the importance of this change in the transition to the role of family care-giver. After all, once a physical move has been made which brings the older relative needing some help and the family care-giver closer together geographically, the nature and content of their relationship will also change, emphasising the commitment made on the part of the care-giver (Stueve and O'Donnell 1989; Stoller *et al.* 1992).

A 49-year-old woman, who was formerly a long-distance care-giver providing regular but intermittent assistance to her elderly mother (average of two hours a week), stated: 'We no longer have to drive 250 miles several times a year to assist with her household and other needs'.

A similar sentiment is expressed by another woman (55 years old, helping her 89-year-old mother) who has made the commitment to take on more direct eldercare. In the transition her elderly mother has moved thousands of miles. This woman, a resident of British Columbia, stated: 'My mother moved from Quebec to Dawson Creek, British Columbia, which has made our family more directly involved'.

These two comments from case study participants give some indication of the magnitude of the changes families undergo when the commitment is made by an adult child, in these two examples (and quite typically) by daughters, to take on the role of eldercare provider. In the transition, their family geographies have changed considerably! Not all moves are of such great distance of course, but still illustrate the relationship between reducing the time–space of care-giving and the clear transition into the role of family eldercare provider. The following comment by a 54-year-old male care-giver, about caring for his 83-year-old mother, illustrates this point: 'Now I have Mom here to look after. It is easier to provide care (now) than driving to the other end of the city'.

Insights into the transition to eldercare

Insight from a time–space perspective

From the data presented here, a picture develops of women eldercare providers not only spending more time in care-giving and associated travel, taking on more eldercare tasks (despite responsibilities to young children) and also being more active in residential relocation – in manipulating their personal eldercare geographies – in order to provide family-based assistance for their elderly relatives. Male care-givers appear to feel the limiting effects of time-distance more, and are much less engaged in residential relocation. Therefore, like the geographies of women with young children (Dyck 1990), the time-geographies of women caring for elderly relatives also reveal a distinctive 'time–space zoning' (Rose 1993). This zoning reproduces a (patriarchal) social order within the activities of everyday family life.

However, the women in this case study clearly show that they are actively (re)negotiating the time and space constraints of their eldercare geographies. As we have seen, their greater engagement in eldercare provision includes a greater likelihood of arranging for an elderly relative to live nearby. There are costs to be borne by women at this stage of the life-course in their transition into the role of eldercare provider. Comments by some of the female care-givers in this case study indicate that the time spent providing eldercare is time taken from their marital and other family relationships which may be strained by the forfeiture of this personal time. This continuance of the care-giver role for middle-aged women may be particularly problematic in situations where spouses were expecting increases in personal time spent together after the 'launching' of their own children (Googins 1991).

At the same time however, women who undertake residential relocation of their elderly relatives report that this was a positive move which allowed them to spend the time, and give the care and assistance, that their family members needed. Residential relocation, then, appears to be a good indicator of a smooth passage into the role of family care-giver, making it easier to fulfil this commitment and at the same time perhaps minimising some of the disruption in their other work and family relationships that frequent long-distance travel may have caused.

Insights from a life transition perspective

As stated earlier, the major life-stage transitions, such as adolescence, marriage and the birth of a child, are generally marked by some form of ceremony or celebration. These ceremonies or 'rites of passage' assist in the establishment of the new roles and responsibilities that go along with the transition into a new social status. Understanding the time spent on care-giving for elderly family members, and especially residential relocation, as adjustments to eldercare geographies, and therefore as tangible points of transition into the mid- to later-life role of family eldercare provider, draws much-needed attention to the experiences of this part of the life-course.

Importantly, when we realise and accept that the assumption of care for elderly relatives is a major life transition, filled with complex adjustments and similar in many respects to earlier transitions in the life-course, we are forced to consider why the transition into eldercare is routinely *not* marked with any sort of ceremony or rite. It appears that this is one more aspect of daily life in Western patriarchal cultures which reflects inherent gender inequalities, and pervasive ageism (Abel 1989; Friedan 1991; Glendon 1997). If as a society we can do this, perhaps soon we will see something analogous to the 'crone-ing ceremonies' (Pinkola Estes 1997) which have become popular among some women wishing to celebrate the freedom and self-knowledge of the post-menopause life stage, develop to celebrate the caring and commitment to another person that mark the transition into eldercare.

Placing eldercare in the list of life-stage transitions also allows us to gain a more complete view of the role of gender and geography and their interconnections in the assignment of care to dependent family members. As noted earlier in this chapter, much research attention has been focused upon the relationships between gender, geography and child care (Dyck 1990; Gregson and Lowe 1995). Adding the transition into eldercare into the picture allows us to develop our understanding of how the role of care-giver may be differentially influenced by gender roles and by family geographies, as they evolve over the life-course (Katz and Monk 1993). This more longitudinal approach to understanding personal geographies presents itself as an intriguing and potentially fruitful means of gaining insight into the ways we are all 'embodied' in our geography.

Notes

1 CARNET was a multi-university, inter-disciplinary research network funded by the Government of Canada under the Networks of Centres of Excellence Program.

2 These categories represent natural breaks in the time-distance data. Results were not found to be sensitive to the time-distance categories.

3 Note that the figures presented are those for respondents who answered 'yes' to the consideration or arrangement of the specific residential relocation option.

References

Abel, E.K. (1989) 'Family care of the frail elderly: framing an agenda for change', *Women's Studies Quarterly* 1–2: 75–86.

Abel, E.K. and Nelson, M.K. (1990) *Circles of Care: Work and Identity in Women's Lives*, Philadelphia: Temple University Press.

Berman, C. (1996) 'Parenting your parents', *American Health* October: 47–49.

Brody, E. (1990) *Women in the Middle: Their Parent-care Years*, New York: Springer.

Chappell, N. (1991) 'Living arrangements and sources of caregiving', *Journals of Gerontology* 46: 1–8.

Chapple, E. and Coon, C.S. (1942) *Principles of Anthropology*, New York: H. Holt and Co.

Daly, K. (1996) *Families & Time: Keeping Pace in a Hurried Culture*, London: Sage.

Davies, K. (1994) 'The tensions between process time and clock time in care work: the example of day nurseries', *Time and Society* 3: 276–303.

Duffy, A., Mandell, N. and Pupo, N. (1989) *Few Choices: Women, Work and Family*, Toronto: Garamond Press.

Dyck, I. (1990) 'Space, time and renegotiating motherhood: an exploration of the domestic workplace', *Environment and Planning D: Society and Space* 8: 459–483.

Finley, N.J. (1989) 'Theories of family labour as applied to gender differences in care-giving for elderly parents', *Journal of Marriage and the Family* 51: 79–86.

Foot, D. and Stoffman, D. (1996) *Boom, Bust & Echo: How to Profit from the Coming Demographic Shift*, Toronto: Macfarlane, Walter and Ross.

Friedan, B. (1991) *The Second-stage*, New York: Bantam Doubleday Dell Publishing Group.

Glendon, M. (1997) 'Feminism and the family: an indissoluble marriage', *Commonweal* 124, 3: 11–16.

Googins, B.K. (1991) *Work/Family Conflicts: Private Lives – Public Responses*, New York: Auburn House.

Gregson, N. and Lowe, M. (1995) '"Home"-making: on the spatiality of daily social reproduction in contemporary middle-class Britain', *Transactions of the Institute of British Geographers* 20: 224–235.

Hallman, B.C. (1997) 'The spatiality of eldercare: towards a gendered geography of the aging family', unpublished Ph.D. thesis, University of Guelph, Ontario, Canada.

Hallman, B.C. and Joseph, A.E. (forthcoming) 'Getting there: mapping the gendered geography of caregiving to elderly relatives', *Canadian Journal on Aging*.

Hanson, S. and Pratt, G. (1995) *Gender, Work and Space*, London: Routledge.

Joseph, A. E. and Hallman, B.C. (1996) 'Caught in the triangle: the influence of home, work and elder location on work–family balance', *Canadian Journal on Aging* 15, 3: 393–412.

—— (1998) 'Over the hill and far away: distance as a barrier to the provision of assistance to elderly relatives', *Social Science and Medicine* 46: 631–639.

Katz, C. and Monk, J. (1993) 'Making connections: space, place and the life course', in C. Katz and J. Monk (eds) *Full Circles: Geographies of Women over the Life Course*, London: Routledge, 264–278.

Kaye, L.W. and Applegate, J. S. (1990) *Men as Caregivers to the Elderly: Understanding and Aiding Unrecognized Family Support*, Lexington, Kentucky: Lexington Books.

Knox, P. and Marston, S. (1998) *Human Geography: Places and Regions in Global Context*, New York: Prentice Hall.

Kobayashi, A., Peake, L., Benenson, H. and Pickles, K. (1994) 'Introduction: placing women and work', in A. Kobayashi (ed.) *Women, Work and Place*, Montreal & Kingston: McGill-Queen's University Press, xi–xiv.

Litwak, E. and Kulis, S. (1987) 'Technology, proximity and measures of kin support', *Journal of Marriage and the Family* 4: 649–661.

Massey, D. (1994) *Space, Place and Gender*, Cambridge: Polity Press.

Pinkola Estes, C. (1997) *Women who Run with Wolves: Myths and Stories about the Wild Woman Archetype*, New York: Ballantine Books.

Rose, G. (1993) *Feminism & Geography: The Limits of Geographical Knowledge*, Minneapolis: University of Minnesota Press.

Schoonover, C.B., Brody, E., Hoffman, C. and Kleban, M.H. (1988) 'Parent care and geographically distant children', *Research on Aging* 10: 472–492.

Silverstein, M. and Litwak, E. (1993) 'A task-specific typology of intergenerational family structure in later life', *The Gerontologist* 33: 258–264.

Stoller, E.P., Forster, L.E. and Duniho, T.S. (1992) 'Systems of parent care within sibling networks', *Research on Aging* 14: 28-49.

Stueve, A. and O'Donnell, L. (1989) 'Interactions between women and their elderly parents', *Research on Aging* 11: 331–353.

Van Gennep, A. (1960) *The Rites of Passage*, trans. M.B. Vizedom and G.L. Caffee, London: Routledge and Kegan Paul. First published 1909, *Les Rites de Passage*, Paris: Noury.

14

SINGAPORE'S WIDOWS AND WIDOWERS

Back to the heart of the family

Peggy Teo (Singapore)

In this chapter, I explore widowhood in an Asian context, using Singapore as an example. I discuss how the widowed cope and how space is used in this process. Singapore is an interesting case study. Even though it is a newly industrialised country with a GNP of over US$26,453 per capita (Department of Statistics 1996: 2), its social values are still essentially conservative and the widowed are 'incorporated' into the household so that the transition phase as suggested by van Gennep (1960) is fairly short. This is a situation which lies in stark contrast to the West. Most of the literature on widowhood in the West portray the experience as a traumatic and burdensome one (see, e.g., Lopata 1973, 1996). In the UK and USA, a high proportion of those who live alone are widowed and rely on friends for companionship and affirmation of self-worth. They are considered a vulnerable group whose level of poverty is likely, although not always, high (Arber and Ginn 1991: 171; Malveaux 1993). As a significant rite of passage, these depictions of widows and widowers lead us to the conclusion that associated changes in life circumstances result in new social positions which are not always favourable.

A complete study of widowhood in Singapore needs to include an analysis sensitive to the multiracial, multicultural society that it is. The ethnic composition which includes the predominant Chinese (75 per cent) and the minority Malay (14 per cent), Indian (10 per cent) and Others (1 per cent) convey rich and complex experiences which warrant separate studies for their respective perceptions on widowhood. However, for the purposes of this chapter, I will draw only a general picture, based on the majority Chinese experience.

Two minus one equals zero: identity issues

Widowhood is for many an unpleasant, distressing and painful experience. Even where the spouse may have suffered a long illness and the partner is prepared for death, substantial adjustment is entailed for the person left behind. The most crucial adjustment centres around the widow's/widower's identity. The research of recent decades shows that, in many cases, when the spouse was alive, husband and wife functioned as a couple and were easily integrated with the rest of society, because they constituted a nuclear unit. After the death of one spouse, the widow becomes referred to as the 'fifth wheel' (Lopata 1973: 168 and in DiGiulio 1989: 57) who clogs up normal social life. Lopata (1973, 1996), Silverman (1986) and Rose (1990) provide very detailed elaborations of how widows and widowers are purposely left out of 'normal' social events such as dinner invitations and party functions. They discuss how widows come to the realisation that their own identities are very much tied to that of their husbands. Advantages conferred by marriage include not only companionship, intimacy and material support but also 'someone to negotiate on her behalf in a male-dominated society' (Jerrome 1990: 201). Stevens (1995), Campbell and Silverman (1996) and van den Hoonaard (1997) describe bereavement as a 'shattering of the symmetry' of couples which can strain even the most long-established relationships. In essence, a widow is a social embarrassment who will increasingly become excluded from normal social life. Even among friends who come from the widow's own walk of life (i.e. separate and self-created by the widow), the same crisis occurs.

Thus, in the American and European literature on the psychology of coping with widowhood, festive events such as Christmas and Thanksgiving are singled out as critical events in which widows and widowers must redefine their identity as either a single person, a sibling, a mother/daughter or a kin member, in order to overcome the loneliness which derives from their lack of an identity and thus find a location/place actively to pursue their lives (Stroebe and Stroebe 1993). The more one invests in new relationships or resurrects old ones, the more quickly adjustment will occur (Silverman 1986: 4). Lopata (1996) has a similar argument, i.e. that the more complex the life space of an individual (e.g. a working widow; a widow who has other roles such as parenting or active family life with other family members or relatives), the easier it is for the widow(er) to cope. Lopata suggests that they often find fulfilment in religion, in family, in community work or in friends who are also single. In this last category, there is a significant amount of literature which discusses friendships among people of the same gender. After the death of a spouse, women become closer to women friends and likewise men to men friends (Mugford and Kendig 1986; Jerrome 1993). While men's friendships are based on shared interests and sociability, women's are more intimate and intense and tend to focus on conversation and mutual support (Jerrome 1990). In terms of location/space, most of the men's interactions take place outside of the home while women visit each other's homes.

In the assessment of widow identity, account must be taken of cultural differences in the experience of widowhood. In Hindu societies, a widow is given a defined identity but not necessarily a desirable one, in that her identity as a widow allows society to exploit her labour and usurp her/her spouse's capital assets (Lopata 1987; Patil in Lopata 1996; and see Bremmer and van den Bosch 1995 for a more detailed discussion on Hindu widows in the Indian subcontinent). The Jewish widow is also likely to be married off (Lopata 1996: 15–16) as is a Turkish widow since they are viewed as incapable of caring for themselves, and they carry the identity of a 'dependant' (Bremmer and van den Bosch 1995). Whatever the outcome, the presence of rituals in these societies contrasts significantly with the lack of associated rituals in Western society. The latter condition contributes to the difficulty of adjustment to widowhood in the West (van Gennep 1960; Gorer 1965) and rites of passage are considered necessary as they help the widowed to crystallise their new identity. Rituals that last for days and sometimes years and which centre around obligations to the dead and to the living define for the widowed roles which he/she cannot deny. For them, behaviour changes are legitimated by these expectations followed by integration into their new roles.

The assistance given by ritual to the widow(er)'s new identity does not negate the struggles that the new status entails. A discussion of the identity of widows and widowers must necessarily include the premise that gendered narratives are crucial to the construction of elderly identities. Work shapes self-identity and self-esteem and is regarded as a 'crucial element of citizenship' (Laws 1995: 115) but its value is attached to men and geographically to public spaces. Social and physical reproductive work as defined by Chant and Brydon (1989: 10) is associated mainly with women. This binary division has led Rose to conclude that the domestic/private sphere of the home domain is a 'feminine' space while the realm of the public embodies rationality which is a masculine trait (Rose 1993: 18 and 35). When women were young, wifehood and motherhood acquired primary importance as 'natural' social roles for women and this discouraged them from active participation in the public sphere. This is more so for the current cohort of older women in Singapore as many have never participated in the paid workforce. In Hochschild's *The Second Shift* (1989), women in careers were found to contribute more time and energy to household management and to the role of parenting than men. Hanson and Pratt (1988), who also looked at career women, provided similar findings.

The male–female/private–public divide has immense implications for widowhood. Research indicates that the non-working wife typically provides unpaid labour and support for her husband's career, and that the husband becomes dependent on this labour and support. Should the spouse die, women who have not developed their own spheres, friends and/or colleagues may find they lose their identity because their principal roles as wife and mother are intrinsically linked with their husband's identity, and, furthermore, they had depended on his income for the running of the household (Martin Matthews 1987; Connidis 1989). Even among career women, activities and routines are embedded within patriarchal structures so that their identity is similarly linked to their spouse (Rose 1990). Men often find that if their spouse dies before them, they cannot cope with the daily chores and with the absence of a constant companion. DiGiulio (1989: 52) and Campbell and Silverman (1996) argue that the identity crisis faced by men is not as pervasive as for women.[1] While women use their marriage as an anchor for identity, men rely on other ascriptions such as work in the public domain. For many men, involvement in the public sphere was more intense during their married lives and, unlike their spouses, they often fall back to that sphere to cope. Not having provided care-giving roles and not being accustomed to household management, both of which are confined to the private sphere, widowers face a different set of problems. Theirs is mainly the sense of loss and the inability to deal with the drastic change to their lives. The most likely outcome is that they will find a substitute for their deceased wife. In the USA, more than half (52 per cent) of widowed men marry within 18 months of the death of their spouse (Campbell and Silverman 1996: 20) and in the UK, the remarriage rate for men widowed at 55 or above was 32.4 per cent compared to 6.6 per cent for women in 1988 (Arber and Ginn 1991: 163).

Gender roles established early in life and reinforced throughout adulthood therefore have impacts on the widowed. While the spouse was alive, men did most of the maintenance and repair of the home and payment of bills. Women did most of the cooking and cleaning. While widows face many problems with regard to the maintenance of the home, widowers have similar problems dealing with domestic routines (Silverman 1988). The identities and the spaces associated with the two genders invariably contribute to the complex experience of widowhood. These problems are confronted at a time of exceptional vulnerability. Thus, the sense of deprivation and helplessness when bereaved is accompanied by difficult adjustments in the management of daily life arising from the ideology of gender role complementarity (Arber and Ginn 1991: 161).

Although the private/public divide has restrictive effects for the widowed, older people can carve out physical spaces for themselves (Laws 1994, 1995) and are capable of devising creative strategies to overcome oppression especially in the arena of urban planning. How is the sub-group of the aged comprising widows and widowers involved in this process of resistance? Unfortunately the answer to this question is not encouraging. For widows and widowers, space does not exist for them as a group as it would for 'old people' who constitute an undesired 'other' but who comprise nevertheless, a legitimate group. Unless the widowed form a group who meet on a regular basis, they cannot contest with other potential users. The recognition that widowhood is a transitory phase which ends at death or remarriage (as the last phase of incorporation) gives the widowed an even more precarious identity. Until then, they are subsumed under other larger umbrellas such as the elderly, the institutionalised, etc. Retirement communities which have widows/widowers as their residents may attempt to represent the interests of this group by providing more opportunities for increased interaction between the widows and widowers, but the connotation is negative rather than positive, partly because the interaction seems 'forced'. Even those who were widowed young have to function as 'normal' members of society with no special dispensations such as help with child care. Homes are rarely constructed for 'dysfunctional' families (e.g. single parent/mother; see Winchester 1990), a category to which a widow(er) also belongs. The outcome is that the elderly widowed may well feel that the space they occupy is not legitimately theirs, and may feel obliged to give over this space to others who, they are persuaded to feel, are 'more worthy' than themselves. The lack of fulfilment experienced by the elderly is therefore intrinsically tied to the geographical spaces they occupy and the meanings (or lack of it as in the case of widow and widowers) that these spaces hold.

Widowhood in Singapore

Senior citizens aged 60 and above constituted 9.1 per cent of the population in 1990 (Department of Statistics 1992: 3). There is little in the way of detailed information on the elderly in the census data. To obtain a better picture, I also draw upon the 1995 National Survey of Senior Citizens (NSSC) which was based on a 4,750 sample of population aged 55 and above. Dwelling type and size occupied by the elderly in Singapore is analysed in Table 14.1.

In the 1990 census, 51 per cent of the population aged 60 and above were widowed, of which 54.2 per cent were women (Shantakumar 1994: 27). While 20.9 per cent of men aged 60 and above were widowed, three times more women (65.5 per cent) were in the same status. It was reported in the census that more than half (54.7 per cent) of widows aged 60 and above were economically inactive compared with only 18.1 per cent of widowers (Shantakumar 1994: 28). Given that an overwhelming majority of older women, regardless of marital status, (82.5 per cent) never worked, it is likely that their financial situation is far from comfortable. Although female labour force participation rate has increased over the years, a woman's salary is still considered secondary to a man's and the bias has resulted in lower Central Provident Fund (CPF) earnings

Table 14.1 Type of dwelling of the elderly and of all Singaporeans (%)

	Elderly	*Total population*
Landed property	10.0	7.2
Public housing	80.3	85.2
(1–2 room flats)	(10.6)	(6.0)
(3–5 room flats)	(68.9)	(78.7)
Others	2.8	0.5
Condominium or private flats	0.8	3.3
Others	6.9	4.3

Source: Shantakumar 1994

for them to tide over old age. In the *National Survey on Senior Citizens* 1995, it was documented that while 58.2 per cent of male senior citizens had their own sources of income (savings, stocks and shares, property etc.), only 28.9 per cent of women likewise did.

Eighty-one per cent of the elderly above the age of 60 in Singapore live with their children (Mehta 1997: 48). The Chinese and Indians prefer to co-reside with their sons while the Malays prefer their daughters. The very high proportion of co-residence immediately brings to mind the question of whether widows and widowers benefit from co-residence and whether they cope better under such circumstances. Does co-residence mean that widows do not have to face the same painful identity problems that their Western counterparts deal with?

The paper turns to the 1995 NSSC survey to answer this question. Of the 4,750 sampled in this survey, 2,311 (48.7 per cent) were widowed, of which 79 per cent were female. The survey's statistics confirmed census information on the ethnic distribution of the elderly and on their educational level: in the NSSC, 81.1 per cent of the widowed had never had a formal education and only 14.7 per cent had completed the minimum six years of primary education (Table 14.2).

It is fortunate that 98.8 per cent of the respondents had children still living at the time of the survey as 92.1 per cent were not earning an income. They

Table 14.2 Educational level of the widowed

Education level	%
No formal education	81.1
Completed primary	14.7
Completed secondary	3.3
Completed upper secondary/diploma level	0.6
Completed university	0.3

Source: NSSC survey 1995

were classified as 'dependants'. On average, the median size of households in which the widowed lived was 4.27 persons of which 1.6 were in employment. Nearly half (47.3 per cent) cited themselves as the head of their household and 38.2 per cent cited their son or daughter-in-law as the head (Table 14.3). In terms of who owns the residence in which the widowed lived, nearly half (48.5 per cent) either owned the residence they were staying in or co-owned it with other members in the household (Table 14.3). Effectively, the widowed were still staying in their own homes or homes in which they had a part share. This provides them some sense of security as their abode belonged to them and there was no real need to worry about a secure home space.

Continuing to live in their own home means that the widowed do not have a problem of uprooting, nor of contending with the problem of unfamiliar, new spaces. In the case where they move in with their children, some adjustments are expected. The NSSC does not look into this issue but a survey comprising 148 respondents (Chan 1994/95) found that the elderly often preferred and sought the help of family members in their day-to-day routines such as personal hygiene, meals and house-cleaning. More importantly, the elderly were not passive receivers of care but active negotiators who reciprocate the care by performing tasks such as childminding (in the case where small children are present) and non-fiscal upkeep of the home. The dynamics of the households indicate that some degree of give and take is involved and the elderly cannot expect the right of use of home space. In higher income households where live-in maids are present, the maids' duties often include the care of old people, although not

Table 14.3 Head of household in the residence and tenancy status of the residence

	%
Head of household	
Respondent	47.3
Son/daughter-in-law	38.2
Daughter/son-in-law	12.5
Other relative	1.7
Non-relative	0.3
Tenancy	
Owner is self (i.e. widowed respondent)	48.5
Joint owner with other members	20.8
Owner is other member of household	23.8
Tenanted	6.6
Others	0.3

Source: NSSC survey 1995

solely. The nature of the negotiation between older members in a household and the rest over the use of this space is a relatively unexplored issue in Singapore and more work needs to be done to illuminate this.

Household membership was relatively simple for many of the widowed. Almost two-thirds (57.2 per cent) had married children and/or their spouses in the same household. Grandchildren/great grandchildren were also common household members (53.1 per cent). The other significant membership was unmarried children who still resided with their parents (Table 14.4). As with most Asian countries, unmarried children in Singapore continue to live with their parents until they form a conjugal unit of their own. The fact that most households are made up of immediate members of the family suggests that familial ties play a significant role in the adjustment of the widowed. Having easy access to family members means that problems can be discussed and resolved quickly.[2] The widowed can depend on the family members to tide them over their difficult periods, a hypothesis which was borne out in the survey. Many adult children in Singapore not only provided for the material and financial needs of the widowed but also provided the companionship and camaraderie that widows in Western Europe or North America sought among friends rather than family members. On the island city-state, as high as 98.7 per cent of all respondents replied that they were 'happy' with their living arrangements. This *status quo* may change if the proportion of frail people increases. At present, less than 7 per cent of the elderly in Singapore are non- or semi-ambulant.

For now, the confidence that the widowed in Singapore place on their children manifests itself in two ways: in steady financial support and in social interaction. In the survey, only a very small proportion of the widowed depended on their own sources of income. Less than one-eighth (12.6 per cent) drew out interest/dividends/rents for daily expenses; another 7.1 per cent used income from salaries and own businesses; and another 1.3 per cent from pension. A

Table 14.4 Household membership in the residence

	%
Married children and/or spouses	57.2
Grandchildren/great grandchildren	53.1
Unmarried children	42.5
Maid	6.9
Other relatives	2.9
Brothers/sisters	1.3
Other non-relatives	1.3
Parents/parents-in-law	0.7

Source: NSSC survey 1995

small proportion (1.3 per cent) were also making monthly withdrawals from the Central Provident Fund for daily expenses.[3] Annuities made up less than 1 per cent. In contrast, 87.9 per cent of the respondents depended on their children to provide a cash allowance on a regular basis. All considered, children were still the main income providers, with 79.4 per cent claiming that their children's contributions made up more than three-quarters of their monthly income. The majority (80.9 per cent) felt that their income was adequate to cover all their expenses. Should there be a shortfall of funds, children would be the people to whom the elderly widowed would turn for funds even before they drew out their own savings. This indicates that expectations about inter-generational flows are prevalent and that the filial child is the first source of income; 94.2 per cent of the widowed did not make plans for financial security in their old age because of the self-same expectation that their children would look after them. If they should get sick, 68.5 per cent of them would rely on their children's Medisave[4] funds to pay for medical costs, again before resorting to their own savings (only 11.3 per cent).

Social support for the widowed is immensely strong in Singapore, aided by strong familial ties. In the survey, 84.4 per cent of children who were not staying with their parents continued to keep in contact with their parents at least once a week. For those who were living in the same household as their parents, their social interaction included meals together (69.1 per cent), casual conversation (56.8 per cent), discussion about events concerning the family (46.8 per cent), participation in outings/recreational activities together (24.6 per cent) and looking after grandchildren (16.9 per cent). Over a third (64.9 per cent) of the widowed, men as well as women, participated in household chores and 65.6 per cent said they were consulted by family members on matters of importance such as in decision-making. Their confidantes often included their own children (85.3 per cent), followed by their sons and/or daughters-in-law (2.6 per cent). In a crisis, they would also seek out their own children first.

From the data, the passage into widowhood of the elderly does not seem a stressful one as most of their physical and social needs are met by members of the household who incorporate them quickly into their new social positions. Although widows have to contribute to the upkeep of the household by doing maintenance chores, this helps them to accept their old/new role more easily. Involvement with immediate family members helps to place them within the hierarchy in the home and allows them to settle down to a routine.[5] So well-adjusted are the widowed that 70.2 per cent spurned the Parent Maintenance Act as a necessary recourse, regarding it with some scorn. This bill allows for the state to intervene when children do not support their parents financially.

Case studies

To provide a more rounded picture on widowhood in Singapore, two case studies will be provided.

Mrs K (aged 72)

Mrs K's husband died of a heart ailment when she was less than 35 and she had to bring up her two sons alone. She worked as a secretary in a bank and the long hours she put in eventually led her to an administrative position in the same bank. She continued in the same employ until she retired at age 55.

Mrs K had two main concerns as a widow. Her first was financial. After her husband's death, she sold their car and their house and purchased a smaller public housing flat. She commented that being in the private sector where jobs are not 'iron rice-bowl' (meaning not secure), Mrs K strategised that she needed to keep to the same employer to ensure security for her family members. Although she was offered other positions elsewhere, she turned them down. Mr K's hospital bills had come to quite a bit and Mrs K's concern turned to rejection as a strategy of coping:

> When I saw my husband so sick, I told the doctor that I didn't want to operate. In those days, medicine was not as good as today and there was only a slight chance of him surviving. Let him die, I said . . . no point carrying on . . . Even today I say the same. If I should get sick, I've already told my children to let me die. There's no point in dragging on and wasting all that good hard-earned money. Life must go on.

Her second concern had been her children. She had turned to her own brothers and sisters for help immediately after her husband's death and they had provided her with financial help, child care and emotional support. However, according to Mrs K, she had resolved that this was not going to be 'forever' as she recognised that her siblings 'had their own families to look after.' Moreover, she had a job.

In her social life, her social circle narrowed significantly after her husband's death. She rationalised that her lack of social life was inevitable because:

> [I had] no time. I had to work. My boss was very understanding but sometimes I had to work until 11 o'clock at night, especially when there was a big meeting the next day. He didn't know where everything was . . . My children I let the helper look after. That time was very cheap. I had a Malaysian lady who lived with us until the children were quite big. She

> cooked for them and sent them off to school. She looked after them until I came home. What to do? I had no time for my children's schoolwork . . . my sons did not go university because they had to get work . . . only have a diploma. What free time I had, I spent with them. Sometimes I telephoned my friends and we chatted on the phone but seldom we go out except maybe on Chinese New Year. Nowadays, I've lost contact with them because all are so busy with grandchildren . . . like me.

Since her children married, Mrs K moved in with one of her sons to 'help look after the grandchildren and to keep an eye on the maid.' In fact, both children invited her to live with them and to sell off her own house as 'it didn't serve a purpose. It would be a bother to rent out the house'. She was given the luxury of a room to herself, installed with a television set and air conditioner which she seldom used because of her thriftiness.

For Mrs K, the rite of passage had come suddenly and she had had little time to mourn the loss of her loved one. Instead, she had plunged into her new role as the breadwinner of the family and had used that as her reference point for adjustment. By getting rid of her two major assets (the house and the car), she charted a path of quick recovery which centred on her two children. She relocated the family to a new environment and put most of her efforts into maintaining her coveted place in the public sphere of worklife. That her sons were willing to admit her into their families and to keep her in close touch with their lives ensured that she could continue to have an anchor role to attach to after she retired. Today, Mrs K's activity space is confined to the domestic space. She leads an active life in the running of the household – she is in charge of groceries, daily cooking, supervision of the grandchildren and directing the live-in domestic helper. She does not have a circle of friends of her own and keeps herself busy with her family members. She never remarried. She feared that her children would not 'get along' with a stepfather. Their interests were placed before her own and her own ability to manage gave the need for remarriage a low priority.

Mr T (aged 73)

Mr T was a widower for over twenty years. His marriage had not been a happy one so he 'did not feel the loss'. Nevertheless, there were adjustments he had to make. Mr T retired at age 60 (after his wife died) but continued part-time work until he died. Before his wife died, Mr T spent most of his free time at an all-male clubhouse in which he was a lifelong member, having joined when he was in his twenties. He played mahjong (a Chinese card game) and other card games or watched videos. After his wife died, this rhythm continued with the exception that he spent nearly all of his evenings with his 'companion'.[6]

When his wife died, Mr T took over some of the domestic chores, namely grocery shopping. As some of his adult children were unmarried, his daughter, and eventually a part-time cleaner, took over the tasks that his wife used to do such as cooking and cleaning. He maintained his position as head of the household. His seven children, all adults, gave him an 'allowance' for the maintenance of the house, which was mainly spent on food expenses and the hiring of a part-time cleaner. The rest went to the maintenance of a vehicle for the household. The car was shared with the children even though it was registered in his name.

In Mr T's case, home space remained his unchallenged private space. He did not feel compelled to move out as his unmarried children were still dwelling in the house. In that sense, it was not an empty nest. Although Mr T did not feel maladjusted at home, he lacked a companion whom he could relate to. He had a relationship which lasted many years but Mr T did not even consider marrying his lover. His decision arose out of his fear of alienation from his children. In Singapore, remarriage is not a common occurrence since many of the widowed shift their social importance to family members, particularly immediate ones and count them as their 'inner circle' on whom they can rely. The main difference between Mr T and Mrs K is that Mrs K would not even consider remarriage while Mr T preferred to have companionship but felt that remarriage would be too drastic a step to take as it had the potential of disrupting his working relationship with his inner circle.

Conclusion

Changes in life circumstances for the widowed bring with it an identity crisis. There is a transition in this rite of passage which entails the (re)construction of an individual's identity. The widowed have to deal with the idea that they are suddenly marriageable again or they can choose to remain single and re-establish old identities which were subordinated according to their life-course stage e.g. parent, sibling, worker, children. This transition stage can be viewed as liminal simply by virtue that it is temporary, the final stage being remarriage or death.

In the Singapore scenario, widowhood is also a transitory stage of life requiring adjustment in the definition of identity, the use of space and the meaning that spaces have for individuals. Home space can overnight become divested of meaning and seem empty rather than warm. Where this happens, many of the widowed in Singapore take flight and move into the home of a family member (as did 44.2 per cent of the respondents in the NSSC survey). Alternatively, children move in with them. Mrs K provides a good exemplar of the course that the widowed choose.

That widows and widowers in Singapore are so intimately involved in the lives of their children, facilitated by being in the same lived space, has resulted in the low desire for remarriage. Other concerns such as the welfare of grandchildren and children quickly fill in the gap created by the death of the spouse. Activity space is also filled up by the family members because the widowed engage in social events like outings and recreation with family members. From the case studies cited, it seems that the elderly widowed have expectations concerning their own welfare. The Confucian ethos of filial piety is still strong in Singapore and many parents expect their children to look after them. In both case studies, the filial children offered a home or a steady income to support the widowed. Even where children were not living with their widowed parents, a high proportion initiated contact with their parents.

Returning to van Gennep's schéma of separation, transition and incorporation, it is clear that widowhood involves a re-socialisation process and the creation of new socio-spatial patterns. The lesson that Singapore has to offer is simply that incorporation can come about quickly if family members take an active role in 'adopting' their widowed parents. The adoption need not involve physical incorporation as in the elderly moving into the same lived space as children, but if channels of communication are open, elderly widowed seem to adjust to their new status quickly. To what extent Singapore's case study is relevant to others is questionable on a number of counts. In the first place, the island is small and space is at a premium so that it may be that there are few options but for the elderly to stay with their children. However, the unpopularity of granny flats offered by the public housing authority suggests that staying together is voluntary in Singapore. In the second place, sons are expected to assume the head of the household role after the death of the patriarch, and to care for their widowed parent. Thus, it is not surprising that in modern Singapore, where most households are nuclear units comprising husband, wife and two children, a modified version of the nuclear unit, including a parent, is widely used in official statistics. This indicates the pervasiveness of co-residence in Singapore. These conditions may well render Singapore's circumstances unique. At the same time, this Singaporean study adds to the heterogeneity of experiences of widowhood, and helps to debunk the 'myths, overgeneralisations and stereotypes' (Lopata 1996: 220) which have been used to explain and predict related behaviour.

Acknowledgements

The author thanks Dr Paul Cheung of the Department of Statistics for access to the data used in this analysis.

Notes

1 Not all researchers reach this same conclusion. Feinson (1986) and Lund (1989) found no such gender differences.

2 There is literature which suggests that family members are not the best care-givers to the elderly (e.g. Goldstein *et al.* 1983; Mason 1991). While recognising their important contribution, this paper also draws upon a survey conducted in 1995 on 5200 public housing households in Singapore. The study discovered that Singaporeans liked to live with or near to their parents as it is convenient for meals and child-care (Housing and Development Board 1995).

3 The CPF is Singapore's equivalent of social security. Twenty per cent of a person's monthly salary is put aside for CPF, topped up by another 23 per cent from his/her employer. The money in CPF can be withdrawn when the person retires.

4 Medisave is a forced savings scheme in which a certain proportion of the CPF funds are put aside for hospitalisation bills. An individual may use his/her own Medisave or any family member's CPF to pay for these and certain medical bills.

5 In my own research on leisure activities of the elderly, I found that a large proportion, if not all, of their leisure time is spent on family-centred activities such as shopping, watching television together, gardening, cooking, etc. Women in particular did not venture out on their own and made use of family resources more than elderly men in their pursuit of leisure (Teo 1997).

6 The issue of sex did not emerge in the interviews with either Mr T or Mrs K. Neither were willing to discuss this and would, with embarrassment, say that they were too 'old' to think seriously about it. There are many unspoken issues to uncover, especially in a predominantly Chinese society and of a cohort of people where mistresses were socially acceptable. Unfortunately, no further light can be shed for the moment.

References

Arber, Sara and Ginn, Jay (1991) *Gender and Later Life: A Sociological Analysis of Resources and Constraints*, London: Sage Publications.

Bremmer, Jan and van den Bosch, Lourens (eds) (1995) *Between Poverty and the Pyre: Moments in the History of Widowhood*, London and New York: Routledge.

Campbell, Scott and Silverman, Phyllis (1996) *Widower: When Men are Left Alone*, New York: Baywood Publishing.

Chan, Dulcie Sok Fern (1994/95) Negotiating space for elderly persons in HDB estates, unpublished thesis, Department of Geography, National University of Singapore.

Chant, Sylvia and Brydon, Lynne (1989) 'Introduction: women in the third world: an overview', in Lynne Brydon and Sylvia Chant (eds) *Women in the Third World: Gender Issues in Rural and Urban Areas*, Aldershot: Edward Elgar, 1–46.

Connidis, Ingrid (1989) *Family Ties and Ageing*, Toronto: Butterworth.

Department of Statistics (1992) *Singapore Census of Population 1990: Demographic Characteristics*, Singapore: Department of Statistics.

—— (1996) *Yearbook of Statistics 1996*, Singapore: Department of Statistics.

DiGiulio, Robert (1989) *Beyond Widowhood: From Bereavement to Emergence and Hope*, New York: The Free Press.

Feinson, M.C. (1986) 'Aging widows and widowers: are there mental differences?', *International Journal of Aging and Human Development* 23, 4: 241–255.

Goldstein, M.C., Schuler, S. and Ross, J.L. (1983) 'Social and economic forces affecting intergeneration relations in extended families in a third world country: a cautionary tale from South Asia', *Journal of Gerontology* 38, 6: 716–724.

Gorer, Geoffrey (1965) *Death, Grief and Mourning in Contemporary Britain*, London: Cresset Press.

Hanson, Susan and Pratt, Geraldine (1988) 'Reconceptualizing the links between home and work in urban geography', *Economic Geography* 64: 299–321.

Hochschild, Arlie Russell (1989) *The Second Shift: Working Parents and the Revolution at Home*, New York: Viking Penguin.

Housing and Development Board (1995) *Social Aspects of Public Housing in Singapore: Kinship Ties and Neighbourly Relations*, Singapore: HDB.

Jerrome, Dorothy (1990) 'Intimate relationships', in John Bond and Peter Coleman (eds) *Ageing in Society: An Introduction to Social Gerontology*, London: Sage Publications, 18–208.

—— (1993) *Good Company*, Edinburgh: Edinburgh University Press.

Laws, Glenda (1994) 'Oppression, knowledge and the built environment', *Political Geography* 13, 1: 7–32.

—— (1995) 'Embodiment and emplacement: identities, representation and landscape in Sun City retirement communities', *International Journal of Aging and Human Development* 40, 4: 253–280.

Lopata, Helena Znaniecka (1973) *Widowhood in an American City*, Cambridge, Massachusetts: Schenkman.

—— (ed.) (1987) *Widows: The Middle East, Asia and the Pacific*, Durham, North Carolina: Duke University Press.

—— (1996) *Current Widowhood: Myths and Realities*, Thousand Oaks, California: Sage Publications.

Lund, Dale (ed.) (1989) *Older Bereaved Spouses: Research into Practical Solutions*, New York: Taylor and Francis/Hemisphere.

Malveaux, Julianne (1993) 'Race, poverty, and women's aging', in Jessie Allen and Allan Pifer (eds) *Women on the Frontlines: Meeting the Challenge of an Aging America*, Washington DC: The Urban Institute Press, 167–190.

Martin Matthews, Anne (1987) 'Widowhood as an expectable life event', in Victor Marshall (ed.) *Aging in Canada–Social Perspectives*, Ontario: Fitzhenry and Whiteside, 343–366.

Mason, Karen (1991) 'Family change and support of the elderly in Asia', in United Nations Economic and Social Commission for Asia and the Pacific (ed.) *Population and Ageing in Asia*, Bangkok: ESCAP, 65–79.

Mehta, Kalyani (ed.) (1997) *Untapped Resources: Women in Ageing Societies Across Asia*, Singapore: Times Academic Press.

Mugford, Stephen and Kendig, Hal (1986) 'Social relations: networks and ties', in Hal Kendig (ed.) *Ageing and Families: A Support Networks Perspective*, Boston: Allen and Unwin, 38–59.

National Survey of Senior Citizens in Singapore 1995, Singapore: Ministry of Health, Ministry of Community Development, Department of Statistics, Ministry of Labour and National Council of Social Service.

Rose, Gillian (1993) *Feminism and Geography: The Limits of Geographical Knowledge*, Minneapolis: University of Minnesota Press.

Rose, Xenia (1990) *Widow's Journey: A Return to Living*, London: Souvenir Press.

Shantakumar, Gopal (1994) *The Aged Population of Singapore*, Census of Population Monograph No. 1, Singapore: Department of Statistics.

Silverman, Phyllis (1986) *Widow-to-widow*, New York: Springer Publishing.
—— (1988) 'In search of new selves: accommodating to widowhood', in Lynne Bond and Barry Wagner (eds) *Families in Transition: Primary Prevention Programs that Work*, Newbury Park, California: Sage Publications, 200–220.
Stevens, Nan (1995) 'Gender and adaptation in later life', *Ageing and Society* 15: 37-58.
Stroebe, Margaret and Stroebe, Wolfgang (1993) 'The mortality of bereavement', in Margaret Stroebe, Wolfgang Stroebe and Robert Hansson (eds) *Handbook of Bereavement: Theory, Research and Intervention*, Cambridge: Cambridge University Press, 175–196.
Teo, Peggy (1997) 'Older women and leisure in Singapore', *Ageing and Society* 17: 649–672.
van den Hoonaard, Deborah Kestin (1997) 'Identity foreclosure: women's experiences of widowhood as expressed in autobiographical accounts', *Ageing and Society* 17, 5: 533–551.
Van Gennep, A. (1960) *The Rites of Passage*, trans. M.B. Vizedom and G.L. Caffee, London: Routledge and Kegan Paul. First published in 1909, *Les Rites de Passage*, Paris: Noury.
Winchester, Hilary (1990) 'Women and children last: the poverty and marginalization of one-parent families', *Transactions of the Institute of British Geographers* 15, 1: 70–86.

15

THE BODY AFTER DEATH

Place, tradition and the nation-state in Singapore

Brenda S.A. Yeoh (Singapore)

Along with birth, death claims a central place in many cultures as one of the universal rites of passage. In positing a tripartite structure of rites (namely, separation, transition and incorporation), van Gennep (1960) is of the view that it was the theme of transition rather than that of separation which dominated funeral rites. Not only do these rites mark the transition of the deceased from the world of the living towards the world of the dead, they could also ease the transition for people who have been bereaved, drawing them from the world of the dead towards the world of the living (Littlewood 1993: 75). Indeed, in Chinese culture, death is often treated as an important rite of passage, both for the dead as well as for the living. Watson (1988: 4), for example, argues that 'proper performance of the [funeral] rites . . . was of paramount importance to determining who was and who was not deemed to be fully "Chinese" '. Beyond ushering the deceased into a different world, these rites also reinforce values associated with Chinese social structure such as filial piety and the maintenance of lineage, affirm participants' (both the dead and the living) belonging to their ethnic community, and help restore the living to their everyday worlds.

In late modern societies, it is no longer the case that death is a taboo subject, either within society generally or in sociology (Mellor 1993). There is now also growing interest within anthropological and geographical fields in the material expressions and consequences of death, whether in the form of bodily remains and habitations for the dead (the question of the embodiment of death) or in terms of the disposal of the deceased's assets (the question of inheritance) (Clark 1993).

This chapter is specifically concerned with the interplay of Chinese funerary rites and disposal practices in the city-state of Singapore and with changing landscapes of death. It draws in part from an approach within cultural geography which examines landscapes as socio-political constructions (see Duncan 1990). This means that not only are changes in rites and rituals visibly expressed (and as a consequence naturalised) in the form and architecture of material space, but, conversely, external forces (such as the powers of the state as discussed here) which (re)shape the landscape may also impact on social practices. In other words, on the one hand, places for the dead are socially organised and produced by the living; on the other hand, the form that these habitations take also condition people's norms and experiences.

It has been argued that far from being 'waste' space of purely esoteric interest, 'the habitations of the departed invite inquiry into all manner of social and geographical questions' because the jolting, crisis-laden nature of death allows us to 'peel back . . . the central layers of our community's value system – traditions and axioms rarely examined during the routine stretches of our days and years' (Zelinsky 1994: 29). As Henry James observed, it was 'when life was framed in death that the picture was hung up' (quoted in Warren 1994). A number of strands of work has spawned from this basic starting point. There are, for example, studies which attempt to examine how the location and spatial form of cemeteries are shaped by social and cultural influences such as religious beliefs. For example, Zelinsky (1994) explains the geographic pattern of cemeteries in the USA using a combination of traditional factors such as population density and the cost of land but also taking into account social and cultural factors such as ethnic and religious diversity, regional culture and the degree of modernisation. Focusing on the Chinese cemetery, Lai (1974, 1987) and Knapp (1977) examine the influence of *fengshui* (Chinese geomancy) *vis-à-vis* the traditional concerns of locational theory such as accessibility. Another range of studies examines not so much how culture impinges on cemetery location but considers how cultural and social values may be inferred from a study of burial landscapes. Geosophical studies (a geographical tradition which seeks to reconstruct images of 'other' imagined worlds, see Lowenthal and Bowden (1975)) such as Zelinsky (1975) and Knight (1985) uses cemetery place names and information culled from epitaphs to piece together conceptions of the afterworld. A further group of studies takes as its premise the observation that burial rituals and artefacts are really for the living rather than the dead. In this vein, gravestone iconography, epitaphs and funerary monuments are 'read' as a 'text' to uncover hidden cultural meanings and attitudes towards life and death (Ludwig 1966; Jackson 1967/68; Howett 1977; Nelson and George 1982). Hardly any attention has been given to crematoria and columbaria as modern landscapes of death although Teather's (1998a, 1998b) work on these in the context of cemeteries in Hong Kong goes some way to fill this gap.

My previous work (Yeoh 1991; Yeoh and Tan 1995) has attempted to situate burial landscapes (in the context of Singapore) as *contested* spaces within broader socio-political developments. Following the view that space is not a scientific object removed from ideology and politics but is instead political and strategic (Lefebvre 1977), I have argued that the site, location and morphology of burial spaces are invested with different meanings by different individuals, social groups and the

State. The clash of priorities is often resolved through a complicated process of conflict and negotiation. On the one hand, 'dominant' groups construct the burial landscape as a site of control; on the other hand, other 'subordinate' groups may also use it as a site of resistance to resist exclusionary tactics and to advance their own claims. Beyond the colonial period and with the transition to independence, landscapes of death were again implicated as an important focal point of debates in the developing discourse on nationhood and nation-building.

In the context of the above work, this present chapter looks at the role of the nation-state in the post-independence era in shaping the world of death for the individual through interventions in the landscape, as well as the responses of individuals to these changes.

Chinese burial grounds in colonial Singapore

In traditional Taoist belief, the liberty to select propitious burial sites according to the principles of *fengshui*, and sepulchral veneration, are important elements of ancestor worship. *Fengshui*, or Chinese geomancy, was considered central to the Chinese faith because it was believed that it was possible to site the grave in relation to the configuration of the landscape and the vicinity of watercourses in such a way that benign influences were drawn from the earth and transmitted to the descendants of the deceased. In general, a favoured burial site was one situated at the conjunction between the 'azure dragon' on the left and the 'white tiger' on the right, the former signifying boldly rising 'male' ground and the latter emblematic of softly undulating 'female' ground. Such ground contained an abundant supply of beneficial 'vital breath' which augured well for descendants. On the other hand, flat, monotonous surfaces or landscapes characterised by bold, straight lines such as the presence of a straight line of ridges, watershed, railway embankment, road or water running along a straight course tended to concentrate malign influences and were avoided as burial sites (Lip 1979; Eitel 1985). Once sited according to geomantic principles, both the tomb and its sepulchral boundaries were considered inviolable as any interference with them would spoil the efficacy of the *fengshui* and imperil the welfare and prosperity of living descendants. As a material expression of ancestor worship, the burial site serves to link in inextricable ways the world of the living and the world of the dead. Its centrality in Chinese belief is further strengthened by the fact that the various strands of Chinese religion converge in the institution of ancestor worship: while the rituals pertaining to the burial of the dead and the art of divining in grave site selection have Taoist roots, ancestor veneration is also given support from Confucianist perspectives (Tham 1984: 7). As 'sacred' sites (drawing on multiple meanings of the 'sacred'), burial sites were also traditionally considered by relatives of the dead to be immune from government intervention or other external interference.

In Singapore, by the late nineteenth century, Chinese burial grounds which by then occupied large stretches of land both within and close to the city were a major source of contention between the colonial and municipal authorities and the Chinese community (Yeoh 1991). In particular, with the emerging pressure for Western-style sanitary reform at the turn of the century, Chinese burial grounds were perceived as hazardous to public health. With the advent of modern urban planning in the twentieth century, burial grounds were further represented as a threat to the economic ethics of space management and were viewed as 'major space wasters' (Yeoh and Tan 1995). For a variety of reasons including the lack of alternative means of disposal, the weakness of the legislative machinery to enforce grave removal, and the concerted effort on the part of the Chinese to protect their burial sites, the campaign to stop the proliferation of Chinese burial grounds and to free land for urban planning was largely ineffectual during the colonial era. In the post-war colonial era, the debate over the 'burials question' took on a different dimension. In 1952, a Burials Committee was set up to consider, *inter alia*, the idea of encouraging cremation as an alternative means of disposal, and in line with this, the construction of a municipal crematorium (*Report of the Committee Regarding Burial and Burial Grounds* 1952). This however failed to gain the co-operation of the Chinese community and prior to Singapore's independence in 1965, the vast majority of the Chinese dead (89.8 per cent) were buried with only 10.2 per cent opting for cremation (Tong 1988: 34).

The changing landscape of death: strategies of the nation-state

While only about 10 per cent of the Chinese dead were cremated in the closing years of colonial rule, the number of cremations began to climb steeply from the early 1970s. By 1988, 68.1 per cent of the Chinese dead were cremated while the remaining 31.9 per cent were buried (Tong 1988: 34). In 1993, there were 4,625 (32.4 per cent) burials and 9,669 (67.6 per cent) cremations. In the 1990s, cremation was preferred by four in five of those for whom burial is not required by their religion (i.e. all communities apart from the Muslim, Ahmaddiya Jama'at, Jewish, Parsi and Bahai populations) (*Straits Times* 9 August 1994).

The reasons behind such a major change of funeral rites of passage within such a short span of time are complex ones. Jupp's (1993) study of the rapid passage from burial to cremation as the principal practice in the disposal of the dead in post-war England concluded that a range of factors was significant in affecting social attitudes towards death and funeral practices. In religious terms, the major Christian denominations hold the position that the mode of disposal has no consequences for the after-life. Changing residential patterns, increased social mobility and a decline in forms of communal solidarity have reduced the role of neighbours in funerals and increased the latitude of freedom families have in the mode of disposal. With the institutionalisation of death and disposal, 'death-work' (nursing the terminally ill and mourning the dead) which used to be the responsibility of the family (and particularly the women) has increasingly devolved to hospitals, nursing homes, funeral directors and local authorities. This has also promoted 'simpler' forms of disposal. Most critically, the transfer of control over the spaces for the dead from the Church to local secular authorities has encouraged cremation as the more practical and financially attractive option.

In part, the change from burial to cremation as the main mode of disposal among Chinese Singaporeans is a reflection of the weakening hold of 'traditional' ideas and beliefs concerning death and the after-life. According to Tham (1984: 60-61), three sets of beliefs are particularly salient in the performance of traditional funeral rites: the expression of filial piety and family and clan solidarity (e.g. the ostentatious, collective expression of grief at funerals and the elaboration of customs concerning mourning); the idea of pollution (e.g. the need to shield family members and others against contagion and to perform cleansing rites); and the belief in the continuing potent influence of the dead on the welfare and fortunes of the living (e.g. the burning of paper money and other paraphernalia are intended for the deceased but also done in expectation of reciprocity). Tham's (1984) survey to enquire into the religious beliefs of Chinese Singaporeans shows that in general, while about two-thirds of the sample continue to observe ancestor worship, adherence to ritual is highly variable. He writes that while the central beliefs remain potent and continue to shape ritual behaviour, there is an 'overriding indeterminacy' as 'the practice of rites is dependent on who happens to know something about them and whether their observance complies with "customary" expectations' (Tham 1984: 64). Specifically, the fear of retribution from the dead (an important reason for 'proper' burial and geomantically favourable sites) seems to hold much less sway than before.

An important factor in explaining the decline of ritual practice – thus paving the way for a greater acceptance of cremation as an alternative to burial – is the diminished role that regional, dialect and clan associations play in Chinese social life after independence. Once 'the warp and woof' of Chinese society overseas (T'ien, quoted in Tham 1984: 4), these associations played an important role in the social organisation of death, collecting funeral subscriptions to ensure that Chinese immigrants received proper burial, supervising funeral rites and ritual observances pertaining to the grave, and maintaining clan graves and burial grounds as focal points for clan identity. With independence, however, in order to re-orientate the new citizenry away from the more parochial, ethnically bounded concerns towards acceptance of the nation-state framework, many of the functions of these Chinese voluntary associations were assumed by the government, including control of funeral and burial matters (Yeoh and Tan 1995: 187). By the 1970s, the rhetoric of 'national development' and the release of 'sterilised' land for socio-economic development became the most explicit reasons for closing and clearing privately owned Chinese burial grounds. Such rhetoric was backed by considerable powers, most crucially in the form of the Land Acquisition Act. In the decades following independence, numerous large Chinese burial grounds were acquired and cleared for the purposes of New Town development.

As alternative means of managing the disposal of the dead, the government offered burial space in a state-owned public cemetery complex at Choa Chu Kang (Figure 15.1) in the western part of the island while at the same time making it clear that it considered cremation as the only viable, long-term solution (Yeoh and Tan 1995: 191). The earliest government crematorium, situated on Mount Vernon, started operations in 1962 with only one funeral service hall with about four cases of cremation a week. By 1995, it had three service halls and ten cremators and was averaging twenty-one cremations a day (Tailford 1995). The site also includes a columbarium built in several phases (Figure 15.2), comprising niches either arranged in numbered blocks, within a nine-storey 'pagoda-style' building, or in a two-storey 'church-style' building. Towards the end of the 1970s, the Mount Vernon complex, which was primarily meant for the storage of ashes from recent deaths, could no longer cope with the scale of exhumation projects fuelling the demand for columbarium niches. The Housing and Development Board,

Figure 15.1 Chinese cemetery at the Choa Chu Kang complex

Figure 15.2 Columbarium block at the Mount Vernon complex

the government public housing agency, for example, was then undertaking the exhumation of some 17,000 graves from the Kwang Teck Suah cemetery at Chye Kay Road. A new columbarium with 16,720 niches was built and opened for the storage of cremated remains in 1978. Another crematorium-cum-columbarium complex was also built at Mandai. This commenced operations in 1982, equipped with eight small and four normal size crematoria and a total of 64,370 niches for the storage of cremated remains.[1] In addition, Chinese voluntary associations such as the Pek San Theng association were allowed to build columbaria to house cremated remains of the dead exhumed from clan-owned cemeteries acquired by the government (*Pek San Theng Special Publication* 1988: 118–119) while Taoist and Buddhist temples and Christian churches were also allowed to offer amenities for accommodating cremated remains. Financial incentives were also provided to promote cremation *vis-à-vis* burial. The announcement of adjustments in burial (currently standing at S$700–840 for those for whom burial is not required by religion) and cremation rates (S$80–100, the fee for a single columbarium niche being S$500) are usually accompanied by statements such as the following: 'The Government's policy is to encourage cremation rather than burial in order to conserve land, so cremation and columbarium niche fees are deliberately kept lower than those for burial' (*Straits Times* 14 August 1997).

Opposition to cremation among the Chinese did not assume public form but remained an undercurrent throughout the years. While private burial grounds were still available for the reburial of exhumed remains in small plots even after the government had ceased providing land for reburial, families could choose to purchase private plots for reburying their ancestral remains. However, as available plots in private cemeteries were gradually taken up and these cemeteries closed against further burial, those who objected to cremation had few courses of action open to them. Most were forced to accept cremation, although some chose to seek reburial plots away from Singapore, in neighbouring Malaysia and beyond that, in China. Even in a recent exhumation exercise affecting a large Chinese cemetery at Bulim, there were 'numerous requests', generally permitted, for the exhumed remains to be transported out of Singapore for reburial.[2]

While the move towards cremation was largely enforced by the government's stringent control of land use and circumscription of permissible burial space, strategies of persuasion and negotiation were also in place, mainly to show that cremation was in accord with Chinese religious beliefs or ritual practices. To do this, the state depended on the role played by funerary middlemen, the expert managers of the rites of death and disposal.

Within the Chinese community, the power relations between the funeral specialists, be they exhumation contractors, caretakers of funeral parlours, priests or geomancers, *vis-à-vis* the ordinary Chinese were such that ritual practices were directed by the specialists while the people paid for their expert knowledge on the varied rites of each Chinese dialect group. An informant who was both a caretaker and funeral parlour owner pointed out: 'The government cannot very well say that everyone in Singapore has to be cremated since there is supposed to be religious freedom in Singapore . . . so they ask us [funeral specialists] to promote cremation'.[3]

As they were directly in contact with the Chinese masses and held positions of respect and authority given their 'expert' knowledge, these middlemen were more successful in eroding the distrust of cremation without any semblance of coercion. The same informant, who also ran a coffin-making shop near Kampong San Theng in the early 1970s, related some stories of difficulties with cremation in the initial years, difficulties of a practical nature but of deeper significance. For example, the typical Chinese coffins of that time were not suitable for cremation as they were too thick and the only ones available for the furnace were the 'Catholic religious coffins'. The Chinese objected to the use of the latter because these were designed with the Christian cross prominently positioned on the coffins. The informant had to alter the coffin design into a form acceptable to the religious sensibilities of the Chinese. In his own words, 'When we first tried to alter them it was difficult – the suppliers said they did not have those with the "flowers and grasses" design so we then had to find bronze "lion head" design plates, colour them white and then put them together ourselves'.

These funeral specialists were thus crucial as mediators, being both cognisant of Chinese cultural and religious beliefs and at the same time in a position to modify them in acceptable ways. These middlemen were instrumental in converting the Chinese to the idea of cremation because, as traditional managers of death, they were able to draw on their specialised knowledge of Chinese death practices. Also, unlike state officials, they were insiders holding positions of authority within the community rather than 'others' who distanced themselves from the intricacies of Chinese funerary beliefs. As mediators of funerary practice, they were able to frame Chinese religious beliefs in such a way as to accommodate the shifts in practice in the move from burial to cremation.

Changing landscapes of death: individual negotiations

I now turn to examining the way individuals come to terms with changing landscapes of death in Singapore through the accounts of four Chinese interviewees: Koon Teck, a Taoist in his fifties; Tina, a Christian in her early forties; Li Cheng, an agnostic in her late twenties; and Jenny, who, in her mid-twenties, finds herself 'in between' religions.[4]

Death is often an event of crisis proportions which draws the extended family together in expressing at least some semblance of collective loss and grief. At the same time, the decision to cremate or bury the dead may also become a literal 'bone of contention' (to use the phrase of one of the interviewees) subject to negotiation in the context of family circumstances. For Koon Teck, the cremation of one of his uncles some ten years ago caused rifts in the family clan:

> When my fourth uncle died ten years ago, he never mentioned that he wanted to be cremated but he was the first one to be cremated in our family. His children wanted a simple job and cremation was faster and easier. Then, my father made helluva noise. He said [to his brother's children]: 'Why didn't you ask? They said, 'Ask who? Someone has to make a decision'. As a result, nobody [no other member of the extended family] came to the funeral. I went to help with the funeral but the rest boycotted the family.

Five years later, when Koon Teck's father passed away, there was no question of cremation: 'My mother didn't believe in cremation so we chose burial. She said, "Why should we burn? We Chinese don't dispose of the body in that manner" '.

For Chinese such as Koon Teck's parents, a proper burial is a non-negotiable as it signifies one's ethnicity and culture. When pressed further as to why it is not 'Chinese' to burn, few interviewees were able to articulate their feelings apart from constant recourse to the idea that besides being 'unChinese', 'burning' is 'too painful a way to go'. Jenny, however, seems to be able to elaborate much more fully on these commonly held notions:

> I have a definite preference for burial. It's more whole, I think. I don't think the person really goes away if you don't really burn the person. You feel their lingering presence. And then you put it in an urn and stick it in a wall. It's a bit too practical. Maybe I have very romantic ideas about death in a morbid way, but at the end of the day, you look at death and think about life. Perhaps you would want to be a bit philosophical. You won't be totally practical about these things. I think it is also steeped in our culture. We've always had this impression that in China, people are buried more than not. They don't really get burnt. The only anomaly I've ever heard of are the sky rites of the Tibetan monks. They feed them to the vultures and stuff! But so far, we Chinese traditionally don't burn people, we bury people. In Singapore, we burn them for lack of space.

Despite a definite preference to be 'buried' rather than 'burned', however, Jenny (unlike Koon Teck's parents) echoes another commonly held view in accepting the inevitability of cremation in the context of Singapore: 'There is not much land in Singapore. I can understand why the government is worried. So I think practically, burial is out of the question when my time comes'.

While Jenny explains the aversion to, but inevitable need to accept, cremation and columbaria, Li Cheng focuses on the significance of the grave and suggests the possibility of other scenarios. For Li Cheng, a visit to the grave of her ancestors – her paternal great grandfather and his two wives – in the Hock Eng Seng cemetery (a Hokkien burial ground located in Lorong Panchar and surrounded by prime landed properties in what is today one of the most expensive residential areas in Singapore) was a taken-for-granted event 'every Qing Ming [the annual Chinese festival of grave-sweeping] ever since [she] could remember'. The annual occasion was a whole-day affair involving the entire family clan and made memorable by the difficulty in locating the ancestral grave:

> The cemetery was very overgrown and not well-kept . . . [It was essential] that everyone went together at the same time because it was like all over the place, [the grave was] very hard to find. This uncle who died nearly five years ago was the only guy who knew how to navigate

> through the whole labyrinth of graves . . . It was like, north of the tallest tree if you go straight on the road, and then travel thirty steps till you come to a big trunk and then turn left. But, of course, it never works out because there was like, oh no! there are fifty trees in this place! [The graves] were not numbered and it was impossible to find unless you really remembered that place. So there were years when we lost it completely. So we went there and couldn't find it, and go back quite dejected.

Each member of the family clan was involved in the Qing Ming rituals: from finding the grave, to the actual work of cleaning and grass-cutting, and preparing and making offerings of ritual food:

> My father and my uncles would wear army kind of gear with long sleeves, hats, proper shoes, really hardy clothing, and I would have to do the same, And then we would load up the car with sickles and everything. The full works. And then when we reach there, we put down the food and start cutting the grass. Not just on the omega grave itself but all around it as well. And then we start making the offerings. I never knew what the offerings were supposed to be. I think it was more of a gesture than anything. And my aunt would have gone shopping a few days before to buy the right food. Not that she knew what it was. She would just go to the shop and say 'For offering for the dead'. And they would give her the stuff. My aunt's Catholic, you see, but because she is the wife of the eldest son, it's her duty to co-ordinate the food for these outings.

In 1995, the cemetery was acquired by the Urban Redevelopment Authority and by early 1997, all graves were exhumed and the majority of the remains transferred to columbarium niches in Mandai. Li Cheng explains what the 'moving house' (her own phrase) episode means for her:

> All these years, everything sort of revolved around Lorong Panchar. It was the epicentre, and somehow that sort of spatial centre has been taken away. It's sacred ground that should not be violated because it's a space from which we orientate ourselves all the time . . . Then you suddenly realise, after all those years, you can't go back any more.
>
> I don't want to go to the columbarium any more because the meaning would have been lost. I think it's really stupid to think that these things can be transferred. For me, space is space. I don't think you can transfer

> something from one space to another space and think that everything will be the same . . . That trip to that very first epicentre, it is important at the subconscious level. The fact that it [the columbarium niche] is not the original burial spot makes all the difference although you can pretend it doesn't make a difference. The actual physical structure [of the grave], the actual trip of getting lost in the woods and then finding it, navigating by trees and all that. That is what gives the place significance. In the columbarium, it's quite easy [to find the niche]. There are block numbers and all that. It's not a problem, And no longer an adventure. And then it all becomes purely ceremonial.

The whole exhumation and 'moving house' experience has ironically confirmed Li Cheng's own personal choice in favour of cremation, partly because this is a 'safer' option in the Singapore context as burial may lead to exhumation. This does not mean that she accepts the state's rhetoric that exhumation is justifiable and cremation preferable to release 'sterilised' land or to conserve 'scarce' land:

> I suppose it [the loss of the ancestral grave] made me decide that I am not willing to pay the price for progress. In the end, what is it all for? More land so that there could be a better standard of living? In the end, if there ever was a referendum on this issue, I know how I would vote . . . To say that the cemetery is dead and has got no significance is a stupid reason.

Li Cheng attributes the general lack of resistance to the exhumation of burial grounds and the 'easy' acceptance of cremation and columbaria among Singaporeans to what she sees as the people's abdication of responsibility in matters of life and death (literally) to the state:

> I asked my aunt, 'So do you think they will be angry?' I didn't specify who the 'they' were, but she knew immediately. And then she just sort of lapsed back into the typical Singaporean response of 'But it's beyond us! The government wants it, so what can we do?' A sort of absolving of responsibility . . . While many people are afraid that if you don't pay your respects properly [i.e. providing a proper burial and maintaining the grave], 'they' will not give you good luck for future generations, they also think that if you are doing as much as you can up to a point and within all the constraints, you will still be rewarded with good luck. I think that is how most people reconcile themselves to cremating or exhuming their dead.

While Li Cheng prefers cremation as the 'safer' option, she is also adamant that death and disposal must be 'space-locked'. By this she means that she would still want to claim a space in death, a space whose 'meaning' is vested both in its beauty as well as permanence (for her, cemeteries in Singapore only give an illusion of permanence since they are not immune from exhumation while columbaria are artificial and unattractive simulations which fail to match both the complexity and significance of the Chinese burial landscape):

> For myself, I am going to be cremated and maybe scattered in some beautiful landscape somewhere that doesn't change much, some parkland or something, possibly abroad. I don't like the Singapore landscape at all because space here doesn't have any value . . . but I do want to be scattered in a particular space where that space can still be a meaningful space.

It must be remembered that the general acceptance of cremation and the columbarium landscape does not signal the end of traditional Chinese funerary rites. For example, the discourse on *fengshui* has been resurrected in relation to the siting of cremation urns in columbaria. In 1983, the government had to change its previous practice of allocating niches at the Mount Vernon complex sequentially using serial numbers and to allow for some degree of free selection for an additional fee, 'following requests from families, some [of whom] would keep ashes until they can get a niche of their choice' (*Straits Times* 23 December 1983). The upper rows at eye-level are usually preferred to the lower two which are perceived to be 'unfavourable' as these niches are in danger of being touched by sweeping brooms as well as being exposed to dust and dirt. Instances of people consulting professional geomancers to determine the most favourable niches to site cremation urns are also not uncommon (*Straits Times* 9 April 1986). Even without professional consultants, many Chinese will attempt to adapt geomantic considerations as a means of individualising the selected niche, as seen in the care Tina's father took in choosing an appropriate niche as a resting place for her mother's cremated remains:

> My father and I walked all round Mount Vernon. In those days [1985], there were only the slab-like structures, some with a little roof on the top. Neither the pagoda nor the church-like structure was there. My father chose a niche on what he considered to be *fengshui*. He first chose a block that didn't have something else blocking it. It wasn't blocked by another block, and it faced an open area, greenery. Wide open space, a field basically. And in the horizon, you can see greenery, trees basically . . . This was important to him because at our apartment [high-rise block], there was also a good view, a view of the sea. So he wanted the same thing for my mother.

Conclusion

Over the years, the statistics charting the relative importance of cremation *vis-à-vis* burial, the increasing familiarity of columbarium landscapes and the lack of organised protest against cremation clearly point to the success of government policies promoting the 'space-saving' means of disposal as an alternative to the traditional but 'space-wasting' burial. Hidden from view are the constant negotiations, often privatised within the family, which represent individual strategies to come to terms with a particular 'choice' of mode of disposal within 'constraints' which are clearly written into both government regulation and rhetoric. A minority of Chinese (some 20 per cent according to statistics) still hold on to traditional beliefs in burial and resist even considering cremation as an option, justifying their choice on grounds of 'culture'. Others clearly view the shift from burial to cremation, from cemeteries to columbaria, with a sense of loss, whether this is expressed in terms of a deterioration in rites which signified the strength of the family clan; of the disappearance of an 'original' landscape sedimented with the notion of 'going back to one's roots' (either traced to an actual ancestral grave or imaginatively to China as the 'home' of ancestors); or of a dilution of 'symbolic meaning', where such meaning can only reside fully in traditional ways of doing things which have evolved over time and across space, and which cannot be transplanted to quick-fix, human-designed landscapes.

In general, however, these sentiments do not override the acceptance of either the practicality or the inevitability of cremation as the 'norm' for the future. Indeed, cremation is often described by interviewees as the 'faster', 'easier', 'cleaner' or 'simpler' mode of disposal which accords with the dictates of 'modern times'. More crucially, however, many interviewees not only accept but support the government's 'land scarcity' argument. Alternatively, some fear that no burial space will survive the inexorable logic of scarcity, and that burial today will only lead to exhumation, and eventual cremation of remains, in the future. In the words of one of the interviewees, 'Why get buried today to be burned tomorrow? It is like dying twice!'

While some interviewees lament the loss of 'meaning' with the shift to cremation, it is also clear that individuals continue to devise strategies to conserve some of this 'meaning'. The columbarium still serves as a material landscape which, with various adjustments, support the practice of ancestor worship.[5] Traditional beliefs such as *fengshui* can also be re-tailored to suit the columbarium landscape. Alongside considerable change in ritual practices, there is both an inflection and at the same time resilience of meaning as individuals adapt and adjust deeply held beliefs *vis-à-vis* newly constituted landscapes of death.

Notes

1 Information given by the Land Clearance Unit, Housing and Development Board, 20 September 1997.
2 Information given by the Land Clearance Unit, Housing and Development Board, 20 September 1997.
3 Personal interview, 20 July 1992.
4 These four respondents were interviewed in 1997 as part of a larger project on changing landscapes of death in Singapore involving fifteen in-depth interviews with a range of Chinese Singaporeans of different sex, generation and level of professional involvement with the conduct of Chinese funeral rites. Pseudonyms are used to ensure anonymity.
5 Essentially, unlike the grave, there is no individual space in front of each niche for the performance of rites of worship. Instead, these rites (e.g. the burning of joss paper) take place in adjacent communal spaces (e.g. drums for joss paper are placed next to each block of niches).

References

Clark, D. (1993) 'Introduction', in D. Clark (ed.) *The Sociology of Death: Theory, Culture and Practice*, Oxford: Blackwell Publishers/The Sociological Review, 3–8.

Duncan, J.S. (1990) *The City as Text: The Politics of Landscape in the Kandyan Kingdom*, Cambridge: Cambridge University Press.

Eitel, E.J. (1985) *Feng-Shui*, Singapore: Graham Brash.

Howett, C. (1977) 'Living landscapes for the dead', *Landscape* 21: 9–17.

Jackson, J.B. (1967/68) 'From monument to place', *Landscape* 17: 22–26.

Jupp, P. (1993) 'Cremation or burial? Contemporary choice in city and village', in D. Clark (ed.) *The Sociology of Death: Theory, Culture and Practice,* Oxford: Blackwell Publishers/The Sociological Review, 169–197.

Knapp, R.G. (1977) 'The changing landscape of the Chinese cemetery', *The China Geographer* 8: 1–14.

Knight, D.B. (1985) 'Commentary: perceptions of landscapes in heaven', *Journal of Cultural Geography* 6, 1: 127–140.

Lai, D.C-Y. (1974) 'A *fengshui* model as a locational index', *Annals of the Association of American Geographers* 64, 4: 506–513.

—— (1987) 'The Chinese cemetery in Victoria', *BC Studies* 95: 24–42.

Lefebvre, H. (1977) 'Reflections on the politics of space', in R. Peet (ed.) *Radical Geography: Alternative Viewpoints on Contemporary Social Issues*, London: Methuen, 339–352.

Lip, E. (1979) *Chinese Geomancy*, Singapore: Times Books International.

Littlewood, J. (1993) 'The denial of death and rites of passage in contemporary societies', in D. Clark (ed.) *The Sociology of Death: Theory, Culture and Practice*, Oxford: Blackwell Publishers/The Sociological Review, 69–84.

Lowenthal, D. and Bowden, M.J. (eds) (1975) *Geographies of the Mind*, Oxford: Oxford University Press.

Ludwig, A.I. (1966) *Graven Images: New England Stonecarving and its Symbols, 1650–1815*, Middletown: Wesleyan University Press.

Mellor, P.A. (1993) 'Death in high modernity: the contemporary presence and absence of death', in D. Clark (ed.) *The Sociology of Death: Theory, Culture and Practice*, Oxford: Blackwell Publishers/The Sociological Review, 11–30.
Nelson, M.A. and George, D.H. (1982) 'Grinning skulls, smiling cherubs, bitter words', *Journal of Popular Culture* 15, 4: 163–174.
Pek San Theng Special Publication (1988) Singapore: Pek Sang Theng Association.
Report of the Committee Regarding Burial and Burial Grounds, 1952 (1952) Colony of Singapore: Government Printing Office.
Straits Times, various issues.
Tailford, E.J. (1995) *A Guide to Mount Vernon Complex,* Singapore: Environmental Health Department, Ministry of the Environment. Prepared for internal circulation within the Ministry only.
Teather, E.K. (1998a) 'Themes from complex landscapes: Chinese cemeteries and columbaria in urban Hong Kong', *Australian Geographical Studies* 36, 1: 21–36.
—— (1998b) 'Homes for the ancestors: establishing new traditions of burial through the provision of columbaria in Hong Kong', in R. Freestone (ed.) *The Twentieth Century Urban Planning Experience, Proceedings of the Eighth International Planning History Conference*, Sydney: University of New South Wales, 894–899.
Tham, S.C. (1984) *Religion and Modernization: A Study of Changing Rituals among Singapore's Chinese, Malays, and Indians*, Tokyo: Centre for East Asian Cultural Studies.
Tong, C.K. (1988) *Trends in Traditional Chinese Religion in Singapore*, Singapore: Ministry of Community Development.
Van Gennep, A. (1960) *The Rites of Passage*, trans. M.B. Vizedom and G.L. Caffee, London: Routledge and Kegan Paul. First published in 1909, *Les Rites de Passage*, Paris: Noury.
Warren, J. (1994) 'A strong stomach and flawed material: the making of a trilogy, Singapore, 1870–1940', abstract of paper presented at the Institute of Southeast Asian Studies, Singapore, 12 September.
Watson, J.L. (1988) 'The structure of Chinese funerary rites: elementary forms, ritual sequence and the primacy of performance', in J.L. Watson and E.S. Rawski (eds) *Death Ritual in Late Imperial and Modern China*, Berkeley: University of California Press, 11–19.
Yeoh, B.S.A. (1991) 'The control of "sacred" space: conflicts over the Chinese burial grounds in colonial Singapore, 1880–1930', *Journal of Southeast Asian Studies* 22, 2: 282–311.
Yeoh, B.S.A. and Tan, B.H. (1995) 'The politics of space: changing discourses on Chinese burial grounds in post-war Singapore', *Journal of Historical Geography* 21, 2: 184–201.
Zelinsky, W. (1975) 'Unearthly delights: cemetery names and the map of the changing American afterworld', in D. Lowenthal and M.J. Bowden (eds) *Geographies of the Mind*, Oxford: Oxford University Press, 171–195.
—— (1994) 'Gathering places for America's dead: how many, where, and why?' *Professional Geographer* 46, 1: 29–38.

NAME INDEX

Abbey, S.E. and Farfinkel, P.E. 164
Abel, E.K. 208, 221; and Nelson, M.K. 210, 214, 215, 219
Agnew, J.A. and Duncan, J.S. 11
Aitken, L. and Griffin, G. 138
Aitken, S.C. 106, 112, 115, 118, 124; and Herman, T. 43
Allen, I. *et al*. 197, 201, 202
Anderson, R. 193, 194
Arber, S. and Ginn, J. 224, 227
Aries, P. 27
Arms, S. 94
Aronowitz, S. 44
Australian Play Alliance 35

Bakhtin, M. 60, 61
Balaskas, J. 94
Barile, M. 146
Barrett, M. 160
Barthes, R. 105
Bartlett, H. 197
Bell, D. and Valentine, G. 80, 161
Bellah, R.N. *et al*. 177, 183
Benitez, M. 44
Berman, C. 209
Biggs, S. 193, 196, 197, 198, 203
Blakemore, K. and Boneham, M. 193, 203
Blythe, R. 204
Blyton, E. 31
Bond, J. *et al*. 193, 194
Booth, T. 197
Bordo, S. 159
Boulding, K.E. 8
Brabazon, K. and Disch, R. 204
Braidotti, R. 10, 12
Bremmer, J. and van den Bosch, L. 226
Briggs, P. 195
Brody, E. 208
Brookins, C. 43; and Robinson, T. 43
Brownmiller, S. 126
Buber, M. 14
Burt, M.R. and Estep, R.E. 130
Butler, J. 11
Buttimer, A. 2, 162
Bytheway, B. 193, 195, 196, 203, 204

California Association of Homes and Services for the Aging 200
Campbell, J. 104, 105
Campbell, S. and Silverman, P. 225, 227
Canon, M. 176, 189
Castells, M. 3
Centre of Cultural Risk Research 28, 38
Chambers, I. 188
Chan, D.S.F. 230
Chan Tang, W. 176
Chan, W. 177
Chant, S. and Brydon, L. 226
Chappell, N. 208, 211, 219
Chapple, E. and Coon, C.S. 210
Charmaz, K. 158
Chavkin, W. 87, 88
Children's Environments Research Group 45
Chouinard, V. 7, 144, 152; and Grant, A. 7
Clark, D. 240
Clarke, J. 94
Clough, R. 197
Cobb, E. 31

Commonwealth of Australia 35
Conn, M. and Saegert, S. 45
Connidis, I. 227
Cotterell, J. 45
Craton, M. 61, 76
Crawford, M.P. 22
Cream, J. 172
Cunningham, C.J. 36; *et al.* 30, 32, 33, 35, 39; and Jones, M.A. 31, 32, 36, 38
Cutchin, M. 123

Daly, K. 210
Dancy, J. 23
Davies, K. 209, 211
de Beauvoir, S. 8
Deem, H. and Fitzgibbon, N.P. 86
Del Castillo, A. 105
DeLeon, B. 45
Deleuze, G. 11
Dentler, R.A. and Elkins, C. 44
Derezotes, D. 43, 44
Dewey, J. 123
DiGiulio, R. 225, 227
Dobash, R.E. and Dobash, R.P. 134
Donaghy, B. 32
Dorn, M. and Laws, G. 61
Dorney, E. 44
Doucette, J. 147, 148, 149
Driedger, D. 146, 148
Duffy, A. *et al.* 209, 211
Duncan, J.S. 240
Duncan, N. 9, 133, 161
Dyck, I. 209, 220, 221

Eco, U. 74
Eder, D. and Parker, S. 45
Eifermann, R.R. 30
Eitel, E.J. 242
Elias, N. 161
Erikson, E. 43
Erkut, S. *et al.* 44
Esping-Andersen, G. 194

Fasick, F.A. 44
Featherstone, M. and Hepworth, M. 193, 196
Feinson, M.C. 236
Fennell, G. *et al.* 196, 203
Finch, J. 204
Fincher, R. 172
Findlay, A.M. and Li, F.N.L. 189
Fine, M. 44; *et al.* 54, 55
Finley, N.J. 208
Fiske, J. 60–1
Foot, D. and Stoffman, D. 209
Foucault, M. 9, 10, 159, 197
Friedan, B. 214, 221
Friedberg, M.P. 29

Gardner, C.B. 131
Gibson-Graham, J.K. 93
Giddens, A. 4, 10
Glendon, M. 214, 221
Godby, G. 32
Goffman, E. 197, 204
Golding, W. 104, 123
Goldstein, M.C. *et al.* 237
Goodey, J. 129, 139
Goodman, R.J. and Smith, D.G. 200
Googins, B.K. 209, 210, 215, 220
Gordon, R.A. 199
Gorer, G. 226
Greer, G. 195, 196
Gregory, D. 11–12, 23
Gregson, N. and Lowe, M. 209, 221
Grimes, R. 105
Grosz, E. 5, 6
Grundy, E. and Harrop, 194

Hågerstrand, T. 11
Hall, S. 4, 11, 12, 183, 188
Hallman, B.C. 209; and Joseph, A.E. 213
Hammersley, M. and Atkinson, P. 189
Hanmer, J. and Saunders, S. 126
Hannah, M. 6
Hanson, S. and Pratt, G. 209, 211, 226
Haraway, D. 3, 11, 23
Hardie, E.T.L. 179
Harkness, L. 94
Harper, S. 203
Harry, J. 132
Hart, R.A. 30, 32, 33, 37, 45
Hartless, J.M. *et al.* 128
Hartsock, N. 7, 23
Harvey, D. 2, 3
Hassell, A.J. 32
Hazan, H. 193, 204

Heller, A. 187
Henrard, J.-C. *et al.* 201
Herbert, C.M.H. 131
Herod, A. 64
Hillman, M. 32, 33, 35
Hilton, J.M. and Haldeman, V. 45
Hochschild, A.R. 226
Homans, H. 106, 109
Housing and Development Board 237
Howett, C. 241
Huang, E. and Lawrence, J. 189
Hughes, R. 190
Hugman, R. 193, 194, 196, 197, 198, 200, 201, 205
Huizinga, J. 30

Irigary, L. 11
Israel, P. 146–7

Jackson, J.B. 241
Jackson, P. 74
Jamieson, A. 202
Jarrett, R.L. 44
Jefferys, M. and Thane, P. 205
Jerrome, D. 225–6
Jessup, D.K. 44
Johnson, H. and Sacco, V. 133
Johnston, R.J. *et al.* 23
Jones, M.A. 34, 35
Jorgensen, D. 64
Joseph, A.E. and Hallman, B.C. 209, 210
Jupp, P. 243

Kahn, R. 106
Karsten, L. 30
Katz, C. 43, 45; and Monk, J. 209, 211, 221
Katz Rothman, B. 79
Kaye, L.W. and Applegate, J.S. 208
Kearns, R. 1
Keith, J. 205
Kent, J. 31
Kimball, S.T. 1, 21–2, 23, 78
King, A.D. 12, 14
King, R. 189
King, T. 86
Kirk, W. 8
Kitzinger, S. 80, 106
Knapp, R.G. 241
Knight, D.B. 241
Knox, P. and Marston, S. 213
Kobayashi, A. 159; *et al.* 211
Komaroff, A.L. and Fagioli, L. 163
Koskela, H. 135, 227
Kraan, R.J. *et al.* 202
Kristeva, J. 4–5, 16–17, 89, 93, 98–101, 102

Ladd, F.C. 44
Lai, D.C.Y. 241
Lary, D. 189
Laslett, P. 199
Latimer, J. 198
Lau, S.K. 181; and Kuan, H.C. 181
Laws, G. 161, 196, 197, 199, 200, 226, 227
Lechte, J. 5
Lee, J. 45
Lees, S. 131, 132
Lefebvre, H. 3, 6, 16, 241
Lewis, C. and Pile, S. 60, 61, 76
Li, F.N.L. *et al.* 189
Lip, E. 242
Littlewood, J. 240
Litwak, E. 20; and Kulis, S. 211
Liverani, M.L. 33
Longhurst, R. 79, 81, 89
Lopata, H.Z. 224, 225, 226, 236
Lowenthal, D. and Bowden, M.J. 241
Ludwig, A.I. 241
Lukinbeal, C. and Aitken, S.C. 124
Lund, D. 236
Lunn, H. 31

McCallum, J. 193, 194; and Geiselhart, K. 194, 201, 202
MacCormack, C. 106
McDonald, P. 204
McDonough, A. 152
McDowell, L. 7, 126, 136
McKendrick, J.H. 64
MacKinnon, C. 126, 136
McMaster University 153
McRobbie, A. and Garber, J. 45
Malveaux, J. 224
Man, G. 189
Marcus, S. 3, 93
Marris, P. 4
Martin, E. 94

Mason, K. 237
Massey, D. 208
Matthews, M. 227
Matthews, M.H. 32, 45
Mauldin, T. and Meeks, C. 45
Mead, G. 21
Mead, M. 104, 107
Meade, M. 105
Mehta, K. 229
Mellor, P.A. 240
Michelson, W. and Reed, P. 176
Midol, N. and Broyer, G. 54
Midwinter, E. 137
Minichiello, V. *et al.* 64, 96
Minois, G. 205
Mirrlees-Black, C. *et al.* 133
Moi, T. 5
Moore, R.C. 29, 30, 32, 34, 38, 45
Morris, J. 190
Moss, P. 161; and Dyck, I. 160, 161
Mugford, S. and Kendig, H. 225
Myerhoff, B. 6, 22, 60

Nadel, A. and Woo, A. 181
Nelson, M.A. and George, D.H. 241
Nettleton, B. 30, 35
Newman, S. 197, 201, 204
Newson, J. and Newson, E. 29

Oakley, A. 92
Opie, I. and Opie, P. 30, 31
Orrn, E. 49
Ozanne, E. 201

Paige, K.E. and Paige, J.M. 108
Pain, R. 133, 137, 227
Painter, K. 227
Parker, S. and Parker, H. 129
Pastor, J. *et al.* 44
Peace, S. *et al.* 197, 198, 199, 201, 205
Pearce, J.C. 29
Peters, L. 22
Phillipson, C. 195
Pile, S. 7, 8; and Thrift, N. 11
Pinkola Estes, E.C. 22, 221
Pollock, G. 4, 22
Population Censuses and Surveys 194
Porteous, J.D. 2
Pred, A. 3
Pringle, R. 91, 94, 95
Probyn, E. 9, 20

Ransome, A. 31
Read, P. 2
Relph, E. 2
Riddell, S. 49
Robertson, G. *et al.* 177
Rojek, C. 70
Roper, D. 5
Rosario, L. do 182
Rose, G. 8, 11, 91, 209, 212, 220, 225, 226, 227
Roth, R. 172
Ruddick, S. 118
Russell, D. 126
Russell, H. 30
Rutherford, J. 188

Saegert, S. 45
Said, E.W. 178, 186, 187
Samuels, M.S. 8
Saul, A. 92, 94–5
Schaefer, K.M. 163
Schoenberger, E. 64
Schoonover, C.B. *et al.* 211
Schouten, J.W. 14, 60
Scraton, S. 44, 45
Scull, A. 197, 202
Seamon, D. 8
Seebohm, K. 60, 61, 73
Serres, M. 5
Shakur, S. 43
Shantakumar, G. 228
Shields, R. 60–1, 63, 68, 71
Shilling, C. 161–2
Showalter, E. 164
Sibley, D. 2, 4
Sikes, G. 43
Silverman, P. 225, 227
Silverstein, M. and Litwak, E. 211
Simon, P. 204
Sinclair, I. *et al.* 197, 198
Siu, H. 181
Skeldon, R. 188
Smith, N. 124
Somerville, P. 177, 190

Sontag, S. 195
Spivak, G.C. 160
Spradley, J. 64
Stanko, E.A. 126, 127, 131; and Hobdell, K. 139
Statistics Canada 151
Stattin, H. and Magnusson, D. 45
Steitz, J.A. and Owen, T.P. 44
Stevens, N. 225
Stoller, E.P. *et al.* 211, 219
Stone, S.D. 148, 149; and Doucette, J. 149
Stoney, B. 31
Stow, R. 27, 32
Stretton, H. 35
Stroebe, M. and Stroebe, W. 225
Stueve, A. and O'Donnell, L. 211, 219
Sullivan, M. 86
Swanson, L. 189
Swinney, B. 80

Tailford, E.J. 245
Taylor, J. *et al.* 44, 45
Taylor, R. 92
Teather, E.K. 2, 241
Teo, P. 237
Tham, S.C. 242, 244
Theweleit, K. 7
Thorne, B. 45
Thrift, N. 4, 10, 22
Thurow, L. 194
Toner, J.A. and Kutscher, A.H. 201, 202
Tong, C.K. 243
Torres-Gil, F.M. 201
Tout, K. 203
Tranter, P. 32, 33
Trinh, Minh-ha 186, 187
Tu, W. 183
Tuan, Y.-F. 2
Turner, V.W. 14, 15–16, 18, 189; and Turner, E. 60, 61, 71

Ungerson, C. 202

Valentine, G. 28, 33, 126, 132; and McKendrick, J. 33
Van Andel, J. 34
van den Hoonard, D.K. 225
van der Spek, M. and Noyon, R. 33
Van Gennep, A. 1, 13, 18, 20, 60, 71, 106, 108, 195, 210, 224, 226, 240
Van Staden, F. 45
Van Vliet, W. 45
Varpalotai, A. 44

Walklate, S. 128
Walmsley, D.J. 2
Wang, G. 175
Ward, C. 33
Ward, J.V. 43
Warren, J. 241
Waters, M.C. 43, 45
Wearing, B.M. 44; *et al.* 44
Webber, M.M. 2
Welsh, F. 190
Whelan, A. 94
Wickberg, E. 189
Willcocks, D. *et al.* 196, 197, 198, 201
Williams, H.W. 89
Williams, S. 199
Williams, T. and Kornblum, W. 43
Wilson, B. 20, 21
Winchester, H.P.M. 63, 228
Winnicott, D.W. 43
Wise, S. and Stanley, L. 131, 136
Wolfensberger, W. 198
Wong, S.-L. 175, 181
Woodward, K. 196

Yeoh, B.S.A. 241, 243; and Tan, B.H. 241, 243, 244, 245
Young, F. 108
Young, I. 79, 85, 142, 151

Zelinsky, W. 241

SUBJECT INDEX

ableist space 17, 143–50
activity space 2, 177
adolescence 15–16; and cheerleading 50–4; and gym as gendered/contested space 47–50; and identity 43–5; neighbourhood study 45–7; and sexual harassment 130–3; social/spatial battle lines 54–6
advice on pregnancy: at Birth Expo 80, 82–3; dietary 80; giving of 78–9, 88; male 80–1, 88; need for 79–83; refusing to heed 87–8; on taking it easy 80, 81–2; on ultrasound scan 80–1
alcohol 68, 88

beaches 15, 16, 60–1, 64, 66, 71, 73, 74, 76
birth centre 16–17; experiences of 95–8; and risk 94; as site for the *chora* 98–101, 102; as space between hospital and home 93–5
Birth Exposition 78, 80, 82–3
body 59; as agent 10–11; defined 6; discursive 159; dualisms concerning 162–3; and identity 11–12; and language 93, 102; and mind 8–9; naturalist approaches to 161, 162; and old age 203–4; as out of control 61, 68–71; and parenthood 107, 108; pleated 9, 12; as political field 9–11; social/cultural definition of 159, 161–2; and space 7; violence to 126–39; whose 7
bounded (material) place 2, 177
Brontë, E. 5

Canada: Chinese migrants in 176–7; disabled women in 142–55
Canadian Aging Research Network (CARNET), Work and Eldercare Research Group 212
care-givers 210; and gender 211–12, 213–14; and time devoted to eldercare 213–15; time-space path 216–17; and travel time 212–13
carnival 16, 59, 61, 71, 74, 76
cascade effect 94, 97
Cavill Mall 66, 71–3, 76
cemeteries: in colonial Singapore 242–3; as contested space 241–2; and *fengshui* 242; as hazardous to public health 243; location/spatial form of 241; and Qing Ming (grave-sweeping) festival 249–50; as 'sacred' sites 242; as space wasters 243
Central Provident Fund (CPF, Singapore) 228, 232, 237
child-rearing *see* parenthood
childbirth 16–17; dualisms concerning 92, 93–5, 96–7, 101; experiences of 95–8; gender issues 91, 92, 94–5, 96; geographical issues 91–3; importance/understanding of 107–8; labour ward/home birth 92, 93–5; local focus 92; and mortality rates 92; natural/medical debate 94–5; power/control in 91; practices of 91, 101–2; RHW (Royal Hospital for Women) birth centre 98–101; and risk 94, 97; rituals concerning 106, 107; rituals of 106

childhood *see* middle childhood
children: adolescent vulnerability of 130–3; and care of widows/widowers 229–30, 231–2, 234, 235–6; impact of on employment 115; and parental warnings 129–30; as quintessential victims 128–30; sexual abuse of 128–30
Chinese migrants: *huagong* (labourers) 175; *huaqiao* (sojourners) 175; *huashang* (traders) 175; *huayi* (Chinese with non-Chinese identity) 175; and identity 182–4; and location of home 184–8; return to Hong Kong 177–80; to Toronto 176–7
chora 4–5, 17; birth centre as site for 98–101, 102
chronic fatigue syndrome (CFS) 164
chronic illness 18, 162; combination approach to 169–71; cultural markers of 163, 164–9; and identity 159–60, 161, 171; invisibility of 163, 166; as journey 18, 158; study 163–4; treatment regime 169–71; as unpredictable, uncertain, unstable 169
columbaria *see* death
communities of interest 2
community, changes in 32
community care *see* home/community care
contested space 3
cremation: acceptance of 244, 247, 251–2, 253; encouragement of 243, 245–6, 253; and *fengshui* 252; increase in 243; opposition to 247, 248; and problems with coffins 247–8
crime, fear of 137–8
cultural marker 163; quest for 164–9
cultural oppression 142, 144, 148, 151
culture: Chinese 182–3; and habits of the heart 177–8; and migration 177–80
cyberspace 3

death 19–20, 253; changing landscapes of 248–52; Chinese burial grounds in Singapore 242–3; and columbaria 20, 245–6, 249, 250–1, 252, 253; and control of funeral/burial matters 244–5; and cremation 20, 243, 244, 245–6, 247, 248; and decline in ritual practice 244; and exhumations 245–6, 250–1; and *fengshui* 20, 241, 242, 252, 253; and funeral experts 247–8; and funerary rites/disposal practices 240–1; ideas/beliefs concerning 244, 253; and individual reactions to 248–52; and influences on modes of disposal 243–4; institutionalisation of 244; and proper burial 248–50; as rite of passage 240; and strategies of nation-state 243–8; as taboo subject 240
disabled women 142–3; academic situation of 143–6; activists in Canada 146–50; being out of place 150–3; boundaries/barriers 152, 153–5; and cultural inclusion 151; and disruption of spaces of oppression 153–5; as economic outcasts 151; and embracing of difference 153–5; and employment 150–1, 154; and homophobia 148, 155; humiliation/pain of 145–6; as invisible 145, 153; knowledge concerning 148; legal duties toward 145; living conditions/social practices of 151–2; and multiple paths of self-discovery 154; as out of place 150–3; as outsiders 145, 150, 152, 153; and phenomenon of denial 145; sexual orientation of 147–8, 152; and state assistance 149–50, 152–3; taking up ableist spaces 143–50; and tax 152
DisAbled Women's Network (DAWN) 147–50, 152–3
discourse 3–4, 159; oppositional 203–4; post-modernist 186, 188–9
distance-decay 213
drugs 68, 88

Edinburgh study 127
eldercare 208–9; case study 212–20; in context 211–12; and distance 215–16; and family changes 219–20; and gender 209–11, 217; insights into 220–1; making time/space for 217–20; and shopping 215; strains of 218–19; time devoted to 213–15; time–space perspective 220; transition perspective 221; *see also* Work & Family Survey

essentialism 158–9
existentialism 8

families: and care of the elderly 208–21; and care of widows/widowers 230–2, 237; influences/tensions in 112–14
fatherhood *see* motherhood/fatherhood
feminism 158–9
fengshui see under death

game stage 20–1
gender: and ageing process 195–6; and care-giving 211–12, 213–14; and childbirth 91, 92, 94–5, 96; conflict 15; dominance of boys 47–50; and eldercare 208, 209–11, 217; gym as contested space 47–50; and identity 43–4, 226; and impact of children on employment 115; parental roles 106, 108–9; and play 30–1, 33–4, 37; public/private divide 226–7; and room to be girls 50–4; social/spatial battle lines 54–6; and sport 44–5, 47–50; and widowhood friendships 225–6
geography, discipline of 1
Gold Coast 61–3, 65, 71; Celebration and Control campaign 74–5

home 2, 4; as anywhere and everywhere 187, 188; definitions of 177, 190; and identity 44; immigrant context 177–8, 180; location of 184–8; parental 112; post-modernist discourse on 186, 188–9; as safe environment 112; search for 186–7; as site of violence 133–4
home/community care 201–3; ambiguities in 202; and identity 202
Hong Kong 190; culture 178–9; identity 181–4; returning to 177–80, 189; *see also Xianggangzen*
huaren (person of Chinese race) 182, 183

identity 3–4, 11–12, 101–2, 158–9; contradictory 163; and culture 12, 109; ethnic 181–4; exploration of 43–5; fluctuation of 163; as fluid 71; formation of 98–101; as fractured 168; and gender 43–4, 226; heterosexual 130–3; and home 44; and illness 159–60, 171; multiple 163; of older persons 197–8, 200, 202; and parenthood 105, 109, 121, 123; reassessment of 176; and school 44–5; and self-concept 11, 43–4; unleashing of 67–8; and widowhood 226–8
identity politics 160–1
illness *see* chronic illness
incorporation stage 13–14, 20–1, 23, 106, 236, 240
institutions 10

jouissance 101
journey: illness as 18, 158; migration as 18; as rite of passage 158

Lamaze, F. 106, 124
Land Acquisition Act (Singapore) 245
language, and subjectivity 93, 98–100
lesbians 132–3; disabled 147–8, 154
life crises 1
life-cycle 13–14, 104; transition phase 60–3
liminal space 13–14, 16, 59, 60–3, 71–2; and suspension of time, space, social order 72–3, 75
lived space 92–3

material (bounded) place 2
M.E. *see myalgic encephalomyelitis*
men 7, 8, 51, 56, 79, 80, 81, 83, 105, 108, 109, 115, 118, 132–41, 151, 208, 212–17, 222, 225, 226, 227, 229, 232, 237
metaphors 157–8
middle childhood 15; adult memories of 31, 32; changing needs of 35–8; freedom/independence in 28–9; and issues of play/space 29–31; physical experience of 27–8; and technological/social change 32–5; and urban planning 38–9
migration 175; and home 181–8; and identity 181–4; as journey 18
mind/body dualism 8–9, 102
motherhood/fatherhood 17, 105, 106; acceptance of 107; differences in 'mothering' 118–21; local, body,

day-to-day context 106–7, 108; as morality play 121; *see also* parenthood
Mr Mom 118
myalgic encephalomyelitis (M.E.) 158; contradictory experiences of 168–9; diagnosis 164–6; illness/health circularity 171–2; and quest for cultural marker 164–9; research study 163–4; socio-demographic profiles 165; symptoms 163; treatment regime 169–71

National Action Committee on the Status of Women 150
National Survey on Senior Citizens (NSSC, Singapore) 228–32

old age 18–19; ambiguous nature of 193–4; and eldercare 19, 208–21; and family care 204–5, 230–2; and fear of crime 137–8; gender differences 195–6; health and welfare policy for 194, 196; and home/community care 201–3; image of 194–6; institutional provision for 18–19; needs of 200, 205; oppositional discourse on 203–4; and process of ageing 194–5; and residential care 196–9; and resistance to danger 138–9; and retirement communities 199–201; social life in 233, 234, 236, 237; as social/cultural construction 193; stigmatism of 19, 203–4
Ontarians with Disabilities Act (ODA) 150, 153

Parent Maintenance Act (Singapore) 232
parenthood 17; and birth process 107–8; complexity of 108; and day-to-day parenting 106–7, 108, 123; and employment 114–22; equity in 115; and gender roles 106, 108–9; and identity 121; and influence/space of elders 109–14; optimism concerning 116, 118; as play 118; responsibilities of 118, 120, 121–2; rite of 123; and work of parenting 114–22; *see also* motherhood/fatherhood
passage 1
patriarchy: and control 92; obstetrics 94, 96, 97; and parenthood 122, 123; rite to 106–9
phenomenology 8
place 2; as activity space 177; as bounded locality 177; and play 29–31; as safe 17; separation of 65–6
Plato 4
play 15, 20–1, 29; after-school 33; content 35, 37; experience of 30; and gender 30–1, 33–4, 37; location 35–7, 38–9; and mobility 32–3, 37, 38; safety/excitement conflict 34–5, 38, 38–9; as unstructured 32
Plunket Society 16, 86–7
positions 3
pregnancy 16; and advice 78–83, 87–8; as medical condition 78, 79, 83, 88; research 78–9; as rite of passage 78; and role of husband/partner 78, 83–4, 88; as shared 83–4; and surveillance 78, 86–7, 88; and touching 78, 84–6; and withdrawal from public places 79
preparatory stage 20–1
psychology 8

radical body politics 162–3
rape 3, 93, 126, 134
research: childbirth 95–8; Chinese migrants 175, 176, 177; chronic illness 163–4; Edinburgh study 127; ethical considerations/power relationships 63–4; family 105–6; in-depth interviews 64; participant observation 64; pregnancy 78; recursive interviews 96; surveys 64–5; transitions 63–5
residential care 196–9; and community contact 197–8; and congregation/segregation 198; as institutional 197; and self-identity 197
residential relocation 209–10, 216–17
retirement communities; congregation/segregation by age 200; defined 199; as exclusive 201; and identity 200; needs of 200; patriarchal element 201; positive features of 200

rites of passage 1–2, 210; able/disabled movement 143–5; and adolescence 15–16; and birth 16–17; collective 148–50; and death 19–20, 240; on display 71; and eldercare 210, 221; as form of myth 105; and illness 18, 157–8, 166, 168; interpretation of 104–5; journeys as 158; lack of 104–5; as learning process 20; life-cycle threshold 13–14; and middle childhood 15; and migration 18; need for 21–2; and old age 18–19, 195, 204; as oppressive/marginalised 17–18; and parenthood 17; post-modern 13–20; and pregnancy 16; preparatory, play, game 20–1; and re-positioning 22; and resistance 22; Schoolies Week 59–60; separation, transition, incorporation 13–14, 20–1, 23, 60–3, 236, 240; and social support 14–15; territorial 20, 22–3; understanding of 157–8; widowhood 224, 232, 234, 235
rituals 13, 21, 22, 76, 104, 106, 109, 125, 210, 226, 240, 241, 242, 244, 250, 253, 255
Royal Hospital for Women (RHW) Birth Centre 93–4; as site for the *chora* 98–101, 102

Schoolies Week 21; as celebratory ritual 65–8; as embodied experience 68–71; location 61–3, 71–3; participants 63–5; regulation/control of 59, 73–5, 76; rehabilitation of image 73–5; as rite of passage 59–76; scale of 63
semiotics 99–100
sense of place 2
separation 13–14, 20–1, 23, 106, 236, 240
sex 69, 237
sexual abuse, in childhood 128–30
sexual harassment, female 130–3, 135–7
Singapore: Chinese burial grounds in 242–3; widowhood in 228–32
socialisation process 3–4, 236
space: battle lines for 54–6; discursive 92–3; female 50–4; male 47–50; and old age 203–4, 228; oppressive 17–18; as restrictive 98; signifiance of 93, 99, 102; teenage use of 45–6; transformative 76; unmasking of 17; women's confidence/vulnerabilty in 135; *see also* time/space relationship
space of flows 3
space of place: activity 2–3; *chora* 4–5; debate concerning 2–5; discursive 3; home 4; positional 3
strategic essentialism 160
structure/agency dualism 10–11
subject: and language 93, 98–100; territories of 11
Sun City 199–200

territorial passage 20, 22–3
Third Age 199
threshold, concept of 13–14
time: and childbirth 97–8, 100–1; cyclical 100; as irrelevant 100–1; linear 97–8, 100; parental 120–1
time/space relationship: and care-giving 209, 210, 211–12, 220; *see also* space
transition phase 13–14, 20–1, 22, 60–3, 106; and breaking-out 65–8; and death 240; dichotomies of 108; and eldercare 210–11, 220–1; and influence of elders 109; researching 63–5; and sense of responsibility 69–70; widowhood 235

violence: and adolescence 130–3; and adults 133–7; and children 128–30; Edinburgh study 127; and old age 137–9; perpetrators of 139; public/private differences 133–7; as regulation of female body 126; resistance to 127; sexual 126; towards men 139; women's experiences of 126–39

widowhood 19, 224, 235–6; case studies 233–5; and child care 233; coping with 225; experience of 225; financial concerns 233; and gender friendships 225–6; and household responsibilities 234, 235; and identity 226–8; and living with children 229–30, 231–2, 234, 235–6; public/private divide 226–8; and re-socialisation process 236; and remaining in own home 230, 234–5; and

remarriage 235, 236; as rite of passage 224, 232, 234, 235; in Singapore 228–32; social life 233–4; social support for 232
women: and academic barriers 143–4; as care-givers 208, 213, 214, 215, 220; with chronic illness 157–72; coping strategies 134–5; devaluation of 144; disabled 142–55; experiences of violence 126–39; and feminine behaviour 131–2; and husband/partner violence 133–5; identities of 159–60; and identity politics 160–1; as ill 159–60, 162; immigrant 176; lack of support for 128; non-heterosexual 132–3; sexual abuse of 128–30; sexual harassment of 130–3, 135–7; and spatial confidence/vulnerability 135; and work-place harassment 135–7
Work & Family Survey 212–13; making time/space for eldercare 217–20; participants 212; residential relocation 216–17; time devoted to eldercare 213–15; time-distances/assistance frequency 215–16; and travel time 212–13; *see also* eldercare

Xianggangren (people of Hong Kong) 178–9, 181, 182, 183; *see also* Hong Kong

yuppie flu 164